Wear of Advanced Materials

Wear of Advanced Materials

Edited by
J. Paulo Davim

ISTE

WILEY

First published 2012 in Great Britain and the United States by ISTE Ltd and John Wiley & Sons, Inc.

ISTE Ltd
27-37 St George's Road
London SW19 4EU
UK

www.iste.co.uk

John Wiley & Sons, Inc.
111 River Street
Hoboken, NJ 07030
USA

www.wiley.com

Library of Congress Cataloging-in-Publication Data

Wear of advanced materials / edited by J. Paulo Davim.
p. cm.
Includes bibliographical references and index.
ISBN 978-1-84821-352-4
1. Mechanical wear. 2. Strength of materials. I. Davim, J. Paulo.
TA418.4.W4174 2012
620.1'1292--dc23

2011044639

British Library Cataloguing-in-Publication Data
A CIP record for this book is available from the British Library
ISBN: 978-1-84821-352-4

Printed and bound in Great Britain by CPI Group (UK) Ltd., Croydon, Surrey CR0 4YY

MIX
Paper from responsible sources
FSC® C013604

Table of Contents

Preface xi

Chapter 1. Carbon Fabric-reinforced Polymer Composites and Parameters Controlling Tribological Performance 1
Jayashree BIJWE and Mohit SHARMA

1.1. Introduction to polymeric tribo-composites. 3
1.2. Carbon fibers as reinforcement 6
1.2.1. Classification of carbon fibers 7
1.2.2. Classification of fabric weaves 12
1.3. Carbon fabric-reinforced composites 12
1.3.1. Manufacturing methods to create CFRCs . . . 13
1.3.2. Performance evaluation of composites 14
1.3.3. Tribological properties 14
1.4. Tribo-performance of CFRCs: influential parameters 15
1.4.1. Influence of the processing technique 16
1.4.2. Influence of fabric contents 19
1.4.3. Fabric orientation effect 29
1.4.4. Effect of fabric weave on performance properties 30
1.4.5. Influence of strengthening the fiber matrix interface 33

1.4.6. Influence of the type of polymer used 41
1.4.7. Influence of the molecular weight of a polymer 42
1.5. Concluding remarks 46
1.6. Bibliography. 50
A1.1. Appendix I: Various techniques for developing CFRCs by compression molding 54
A1.1.1. Hand lay-up technique 54
A1.1.2. Impregnation technique 55
A1.1.3. Polymer film technique 55
A1.1.4. Powder prepreg technique 55
A2. Appendix II: Characterization methods for CFRCs 57
A2.1. Physical characterization 57
A2.2. Mechanical properties. 59

Chapter 2. Adhesive Wear Characteristics of Natural Fiber-reinforced Composites 61
Belal F. YOUSIF

2.1. Introduction 62
2.1.1. Why natural fibers? 62
2.1.2. Tribology of polymeric composites based on natural fibers. 63
2.2. Preparation of polyester composites 67
2.2.1. Preparation of FRPC 67
2.2.2. Preparation of palm-oil fibers and PORP composites 69
2.2.3. NaOH treatment 69
2.2.4. Preparation of PORP composites 70
2.3. Specifications of the fibers and composites. 70
2.3.1. Interfacial adhesion of palm-oil fibers 70
2.3.2. Mechanical properties of the composites 74
2.4. Tribo-experimental details. 76
2.4.1. Experimental procedure 78
2.4.2. Examination of worn surfaces 79
2.4.3. Parameters measured 80

2.4.4. Results and discussion 80
2.4.5. Effect of operating parameters. 80
2.4.6. Effect of 6% NaOH treatment 87
2.4.7. Effect of wet and dry contact conditions 89
2.5. Summary 93
2.6. Bibliography 94

Chapter 3. Resistance to Cavitation Erosion: Material Selection 99
Jinjun LU, Zhen LI, Xue GONG, Jiesheng HAN and Junhu MENG

3.1. Cavitation erosion of materials – a brief review 99
3.2. Measuring the wear resistance of a material to cavitation erosionby using a vibratory cavitation erosion apparatus. 101
3.2.1. General view of an ultrasonic vibratory apparatus 101
3.2.2. Determination of the wear resistance of a material to cavitation erosion. 103
3.2.3. Experimental details 105
3.3. Material selection 108
3.3.1. Metal and alloys. 109
3.3.2. Advanced ceramic. 112
3.3.3. Polymer. 113
3.3.4. Comparison 114
3.4. Conclusion 115
3.5. Acknowledgement 116
3.6. Bibliography 116

Chapter 4. Cavitation of Biofuel Applied in the Injection Nozzles of Diesel Engines 119
Hengzhou WO, Xianguo HU, Hu WANG and Yufu XU

4.1. Introduction 120
4.2. General understanding of cavitation erosion . . . 122
4.2.1. Mechanism of cavitation erosion 122

4.2.2. Synergistic effect of cavitation erosion and corrosion 129
4.3. Hydraulic characteristics of cavitation flow 131
4.3.1. Numerical models and validation 133
4.3.2. Effect of boundary pressure on cavitation . . . 133
4.3.3. Effect of nozzle geometry on cavitation 136
4.4. Influence of fuel property on cavitation. 139
4.4.1. Cavitating flow characteristics 140
4.4.2. Variation in the characteristics of dimensionless parameters 142
4.4.3. Effect of fuel properties on cavitation inception 144
4.5. Cavitation erosion of biofuel in the diesel injection nozzle 146
4.5.1. Effect of cavitation erosion on a nozzle 146
4.5.2. Location of cavitation erosion in a nozzle . . . 148
4.5.3. Factors that influence cavitation erosion in nozzles. 151
4.5.4. Effect of biofuel on the erosion of nozzles 154
4.6. Conclusion 155
4.7. Acknowledgments 156
4.8. Bibliography. 157

Chapter 5. Wear and Corrosion Damage of Medical-grade Metals and Alloys. 163
Jae-Joong RYU and Pranav SHROTRIYA

5.1. Introduction 164
5.1.1. Total joint replacements 167
5.1.2. Metal alloys. 169
5.2. Clinical studies and mechanistic investigation into implant failure 173
5.2.1. Wear mechanisms 176
5.2.2. Physiological corrosion of metals 178
5.2.3. Bio-tribo-corrosion at the modular interface 180
5.2.4. Adverse effects due to the creation of wear particles 183

5.3. Residual stress development by rough surface contact. 184
5.3.1. Surface properties of bio-implants 186
5.3.2. Stress-assisted electrochemical dissolution and local corrosion damage 188
5.4. Conclusion 192
5.5. Bibliography 193

List of Authors 197

Index 201

Preface

In general, we currently define wear as the "progressive loss of material from the operating surface of a body occurring as a result of relative motion at the surface". Wear is related to surface interactions, and more specifically to the form of contact due to relative motion. It is important to distinguish between mechanical wear and other processes with similar outcomes. For example, the current definition does not include:

– impact wear, where there is no relative motion;

– cavitation, where the counterbody is a fluid;

– corrosion, where the damage is due to chemical rather than mechanical action.

The progressive loss of material from surface is rarely catastrophic but it does reduce the operating efficiency of equipment, components and structures.

The purpose of this book is to present a collection of examples illustrating the state-of-the-art and research developments into the wear of advanced materials in several applications.

Chapter 1 presents tribological aspects of carbon fabric-reinforced polymer composites.

Chapter 2 covers the adhesive wear characteristics of the natural fibers of reinforced composites.

Chapter 3 contains information on resistance to cavitation (material selection).

Chapter 4 is dedicated to the cavitation of biofuel applied in the injection nozzles of diesel engines.

Finally, in Chapter 5, the wear and corrosion damage of medical-grade metals and alloys is presented.

The present book can be used as a research book for a final undergraduate engineering course (for example into materials, mechanics, etc.) or as the focus of the effect of wear on advanced materials at the postgraduate level. This book can serve also as a useful reference for academics, biomaterials researchers, mechanical and materials engineers, professionals in related spheres working with tribology and advanced materials. The interest in and the use of the topics covered in this book is evident for many important centers of research, laboratories and universities throughout the world. Therefore, it is hoped that this book will encourage and enthuse others to carry out research in this important field of science and engineering.

I would like to pass on my gratitude to ISTE-Wiley for this opportunity to expand the knowledge of others through the use of this book and I thank them for their professional support. Finally, I would like to thank all of the authors who worked on the various chapters for their work on this project.

J. Paulo Davim
University of Aveiro, Portugal
January 2012

Chapter 1

Carbon Fabric-reinforced Polymer Composites and Parameters Controlling Tribological Performance

The inclusion of carbon fiber reinforcement in composites in order to achieve high performance is currently one of the most attractive solutions when encountering increasing demands on the development of materials as a consequence of innovations in technology. Bidirectional reinforcement, as in the case of fabric, is preferable to the use of unidirectional or short fibers because of the possibility of higher specific strength in both directions and the ease of handling reinforcement during processing.

The performance of such composites is a result of various parameters, mainly related to their development and situations in which they are used under selected operating parameters. In the case of tribology, carbon fiber has special importance as reinforcement. This is especially the case in

---

Chapter written by Jayashree BIJWE and Mohit SHARMA.

polymers because of their additional important features, such as lubricity and high thermal conductivity and stability.

The main parameters responsible for the performance of such composites during development are:

– type of matrix and its molecular weight;

– type of carbon fibers (polyacrylonitrile [PAN], pitch etc., or strands, tows, etc.);

– amount of fabric and its weave;

– orientation of fiber/fabric with respect to loading direction;

– fiber–matrix interface;

– processing technique;

– various parameters.

In spite of lot of work reported on carbon fabric-reinforced polymer composites, no in-depth information presenting an overview of such composites is currently available.

This chapter provides a comprehensive review of the parameters of such composites and their influence on performance properties (mechanical and tribological in various wear modes) is presented by our development of a number of composites by varying one parameter at a time. It is concluded that the above-mentioned parameters significantly control the performance of composites. The influence of parameters on tribological properties does, however, depend on the modes of wear that are selected. In this chapter, we show that reinforcement proves significantly beneficial for adhesive and fretting wear situations; whereas in abrasive and erosive situations it proves detrimental.

1.1. Introduction to polymeric tribo-composites

From the tribological point of view, polymers have key features such as self-lubricity, resistance to wear (in dry conditions), corrosion, impact and shocks. Apart from ease in processing of components, they offer quiet operation because of very good damping capabilities. Polymeric tribo-composites also have serious limitations, such as low thermal stability, low strength and deterioration at elevated temperatures. Hence, they are almost invariably used in a composite form.

These composites are used in a variety of tribo-applications, such as ball bearings, cages, bushes, marine equipment, etc., and in load-bearing applications, such as struts, chassis and brackets in automotive and aircraft structures. This is because of their high flexural modulus, compressive strength and high resistance to corrosion [DOS 87]. In the aerospace industry, potential uses of composites containing graphite and carbon fibers include:

– their use as liners for self-aligning plain spherical bearings, cages and braces for ball and roller bearings;

– their use as a seal material for sliding-contact seals, piston rings, valves, bearings in copiers, business machines, space vehicle components, etc. [FUS 88].

Composites contain various constituents, such as fibers, fillers and solid lubricants of various types and sizes. Fibers generally increase load-carrying capacity and strength, and reduce the extent of the interaction of a polymer with the counterface, and hence reduce wear. Fibers are far more resistant to wear than the matrix, and the wear of fiber-reinforced polymers (FRPs) is mainly controlled by fibers. The role of the matrix is to hold the fibers firmly in adverse conditions involving thermal and mechanical stresses. The performance of FRPs depends mainly on the type of fiber/s and matrix, concentration, distribution, aspect ratio, alignment with respect to loading direction, its adhesion to the matrix, processing technology, etc. Fibers with a high aspect ratio (l/r, where l and r are the length and radius of a fiber), have less chance of a concentration of flaws on their surface during loading, which effectively improves the rate of load transferred from the matrix to the fiber and hence the wear resistance of composites, W_R (inverse wear rate) [FRI 86].

$$\sigma_f = 2\tau l r^{-1} + \sigma_m \qquad [1.1]$$

where:

– σ_f is the contact stress;

– σ_m is the compressive stress of the matrix in the composite loaded against a counterface under a load W;

– τ is the tangential stress produced because of the difference in the moduli of a matrix and the fiber.

FRPs are mainly of two types – short fiber-reinforced polymers (SFRPs) and continuous fiber-reinforced polymers (CFRPs). Table 1.1 indicates the range of tribo-potentials and the application areas of such composites, including those of thin-layer composites. In the case of CFRPs, various possibilities exist, such as:

– unidirectionally reinforced with long fibers (UD);

– bidirectionally reinforced with woven or non-woven fabric (BD);

– three- or multi-directionally reinforced with the proper arrangement of fibers/fabrics in three or more directions (TD/multi-D).

Among these, SFRPs are the easiest to manufacture, with a very high production rate through injection molding. However, such SFRPs have comparatively lower tribo-potential properties and strength (Table 1.1), while UD composites have moderate potential.

The manufacturing of composites is not carried out by injection molding, but by compression molding in general. Processing is very difficult in the case of UD composites basically because of the difficulties involved in handling the fibers. BD reinforcement is the most promising because of its multiple advantages, such as its very good strength properties in both directions and ease of fiber handling during processing.

Among the three classes of polymers, known as elastomers, thermosets and thermoplastics; thermosets such as epoxies are the most favoured for manufacturing BD composites with carbon fabric for lightweight construction parts, especially in the aircraft industries basically because they have a very good cost-to-performance ratio. For tribo-components, however, such polymers have not proved the right choice because of lower thermal stability and the higher μ offered by epoxies/thermosets. Instead, thermoplastics have proven a better choice, mainly because of their higher thermal stability, as in the case of specialty polymers such as polyimides, polyetherimide (PEI), polyetheretherketone (PEEK), ppolyethersulfone (PES), polytetrafluoroethylene (PTFE), etc., high damping capacity, better tribo-performance and the possibility of reusing the polymer.

Among the most favored tribo-fibers – glass, carbon and Aramid – glass fibers are the cheapest and are moderately effective in reducing wear but generally affect the μ adversely. Carbon/graphite fibers are the most expensive. They are highly effective in reducing both friction and wear, and also act as thermal conductivity boosters. Aramid fibers, on the other hand, are moderate in cost and effective in reducing wear and sometimes also friction. Thus, in spite of their high cost, carbon fabric is the most favoured reinforcement for composites, including tribo-composites, where performance is the decisive parameter rather than the cost [SOU 05].

Composite material	Tribo-applications	Applicable operating parameters (pv, v and T)	Maximum tribological potential
Neat and SFRPs	Seals, gears, slideways bearings, mild abrasive wear applications, etc.	pv <15 MPa.m/s v <5 m/s, T <250°C	μ >0.03 $K_0 > 10^{-16}$ m³/Nm
CFRPs (UD and BD composites)	Under-water or high-temperature applications, aerospace seals, bearings, etc.	pv <100 MPa.m/s v < 5 m/s T < 320°C	μ > 0.09 $K_0 > 10^{-17}$ m³/Nm
Thin-layer composites with metallic supports	High-pressure applications, pivot bearings etc.	pv < 300 MPa.m/s v <1 m/s T <320°C	μ > 0.06 $K_0 > 10^{-18}$ m³/Nm

Key: p – pressure, v – speed, T– temperature, and K_0 – specific wear rate

Table 1.1. *Tribo-potential of polymers and composites for a variety of applications [FRI 93]*

1.2. Carbon fibers as reinforcement

Carbon fibers (about 5–10 µm in diameter with a density of around 1.78 g/cc) predominantly consist of carbon atoms. They have special properties, such as:

– an exceptionally high tensile strength-to-weight ratio;

– high reinforcing capability;

– very low coefficient of linear thermal expansion, which provides good dimensional stability;

– high fatigue strength;

– high thermal conductivity;

– lubricity;

– wear-reducing capability.

They also have some limitations [MAL 08, DON 96, CHU 94, BUN 88, CHO 93, BUR 99, MOR 05], such as:

– low strain-to-failure values;

– low impact resistance;

– higher electrical conductivity, which may cause a short-circuit in unprotected electrical machinery.

The major areas for the application of carbon fibers are shown in Table 1.2.

The atomic structure of carbon in fiber form is similar to that of graphite (a crystalline material), consisting of sheets of carbon atoms called graphenes arranged in a regular hexagonal pattern.

The sheets are stacked parallel to one another in a regular fashion. The intermolecular forces between the sheets (Van der Waal forces) are relatively weak, giving graphite its soft and brittle characteristics. The crystallographic structure provides the attributes of higher tensile strength and modulus in the direction that is normal to the graphene sheets.

1.2.1. *Classification of carbon fibers*

Carbon fibers are commercially available in varying ranges of tensile modulus (237 to 1,035 GPa). Lower modulus fibers have a lower density, lower cost, higher tensile and compressive strength, and higher strain-to-failure ratios.

Physical strength, specific toughness, light weight	Aerospace, road and marine transport, sporting goods, etc.
High dimensional stability, low coefficient of thermal expansion, low abrasion	Missiles, aircraft brakes, aerospace antenna and support structure, large telescopes, optical benches, waveguides for stable high-frequency (GHz) precision measurement frames, etc.
Good vibration damping, strength and toughness	Audio equipment, loudspeakers for hi-fi equipment, pickup arms, robot arms, etc.
Electrical conductivity	Automobile hoods, novel tooling, casings and bases for electronic equipment, EMI (electromagnetic interference) and RF (radio frequency) shielding, brushes, etc.
Biological inertness and X-ray permeability	Medical applications in prostheses, surgery and X-ray equipment, implants, ligament repair, etc.
Fatigue resistance, self-lubrication, high damping	Dry bearing applications
Chemical inertness, high corrosion resistance	Chemical industry; nuclear field; valves, seals, pump components in process plants, etc.
Electromagnetic properties	Large generator retaining rings, radiological equipment, etc.

Table 1.2. *Properties and applications of carbon fibers ([MAL 08], www.chem.wisc.edu, www.netcomposites.com)*

Carbon fibers are manufactured from synthetic fibers through heating and stretching processes. Depending upon the fiber precursor materials, carbon fibers may be turbostratic or graphitic, or mixed. In turbostratic carbon fibers, the sheets of carbon atoms are haphazardly folded or crumpled. Carbon fibers derived from PAN are turbostratic, whereas carbon fibers derived from the mesophase pitch followed by heat treatment at temperatures exceeding 2,200°C are graphitic. Turbostratic carbon fibers tend to have high tensile strength, whereas heat-treated mesophase-pitch-derived carbon fibers have high Young's modulus and thermal conductivity. Based on precursor fiber material, modulus, strength and final heat treatments, carbon fibers are classified into the categories given below ([MAL 08], www.cytec.com, www.netcomposites.com).

1.2.1.1. *Fiber precursor*

Based upon the precursor material used, the carbon fibers are classified as PAN-based, pitch-based, mesophase pitch-based, isotropic pitch-based, rayon-based, gas-phase grown, etc.

1.2.1.2. *Fiber properties*

Based upon the carbon fiber properties, the classifications include those with:

– ultra-high modulus (UHM) with >450 GPa;

– high modulus (HM) with 350–450 GPa;

– intermediate modulus (IM), with 200–350 GPa;

– low modulus and high tensile (HT), with <100 GPa, tensile strength and >3.0 GPa;

– super-high tensile (SHT) with tensile strength >4.5 GPa.

1.2.1.3. *Final heat treatment*

Based upon the final heat treatment, the carbon fibers are classified as:

– High heat-treatment (HHT) carbon fibers, where the final heat treatment temperature should be >2,000°C and can be associated with a high-modulus type fiber.

– Intermediate heat-treatment (IHT) carbon fibers, where the final heat treatment temperature should be around or above 1,500°C and can be associated with a high-strength type fiber.

– Low heat-treatment (LHT) carbon fibers, where the final heat treatment temperatures are no higher than 1,000°C. These are low-modulus and low-strength materials.

PAN, pitch and rayon are the main raw materials used to produce carbon fibers.

PAN precursors form the basis of the majority of commercially-available carbon fibers and general have higher tensile strengths. These precursors can be thermally modified before decomposition, which allows them to be oxidized and stabilized before the conversion process to carbon fibers, while maintaining the same filamentary configuration. Pitch precursors are based on petroleum asphalt, coal tar and polyvinyl chloride.

Pitches are relatively low in cost and high in carbon yield.

Rayon precursors are derived from cellulosic materials. The high weight loss and low conversion yield to carbon fibers is the main processing disadvantage. Typically only 25% of the initial fiber mass turns into fibers after carbonization, making them more expensive to produce.

The tensile strength of carbon fibers can be increased by hot stretching >2,000°C, during which graphitic planes are aligned in the filament direction. The high modulus of pitch fibers is due to the fact that they are more graphitizable and the shear between parallel planes of the graphitized fiber is easier. These are more sensitive to defects and flaws, and their tensile strength is not as high as that of PAN fibers (in which graphitic basal planes are parallel to the fiber axis) [MAL 08]. The difference in structure of these two types of fibers is shown in Figure 1.1. Table 1.3 gives the typical properties of PAN- and pitch-based carbon fibers.

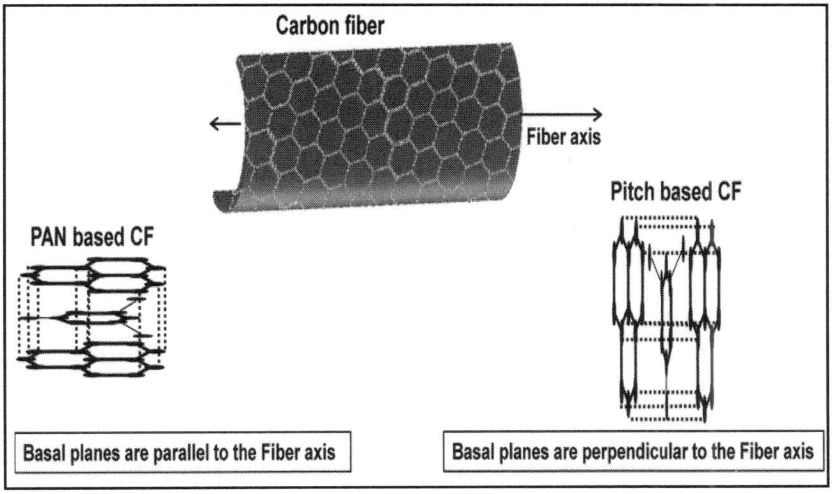

Figure 1.1. *Graphitic structure of a carbon fiber*

Properties/fiber type	PAN-based	Pitch-based
Thermal conductivity (W/mK)	10–100	20–1000
Electric conductivity (S/m)	10^4–10^5	10^5–10^6
Specific heat (at 300 K)	0.17	0.17
Density (g/cc)	1.76	1.90
Tensile strength* (MPa)	434	220
Tensile modulus* (GPa)	235	380

*varies with the type of heat treatment

Table 1.3. *Typical properties of PAN- and pitch-based carbon fibers [KEL 94, MAL 08]*

1.2.2. *Classification of fabric weaves*

Fabrics are woven on looms in a wide variety of weights, weaves and widths. BD woven fabrics provide good strength in the 0° and 90° directions. Variation in *the* placement of fabric orientation *at* different angles can bring *greater* isotropicity in the strength. They also allow *for* faster composite fabrication compared to using two laminates and applying one in each direction. However, woven fabrics provide *a* lower tensile strength than separate laminates because *the* fibers are crimped as they pass over and under one another.

Under tensile loading, these fibers try to straighten out, causing stress within the resin matrix system. These features mainly depend on the weave pattern.

1.2.2.1. *Weave pattern*

Woven fabrics are produced by the interlacing of warp (0°) fibers and weft (90°) fibers in a regular pattern or weave style. The fabric"s integrity is maintained by the mechanical interlocking of the fibers. Drape (the ability of a fabric to conform to a complex surface), surface smoothness and the stability of a fabric are primarily controlled by the weave style. Some common weave styles are discussed in brief in the following section and are depicted in Figure 1.2. General properties of these weaves are shown in Table 1.4.

1.3. Carbon fabric-reinforced composites

The performance of CFRCs mainly depends on the type of matrix and fabric, its amount, weave, orientation with respect to loading direction, fiber-matrix interface, manufacturing technique, etc.

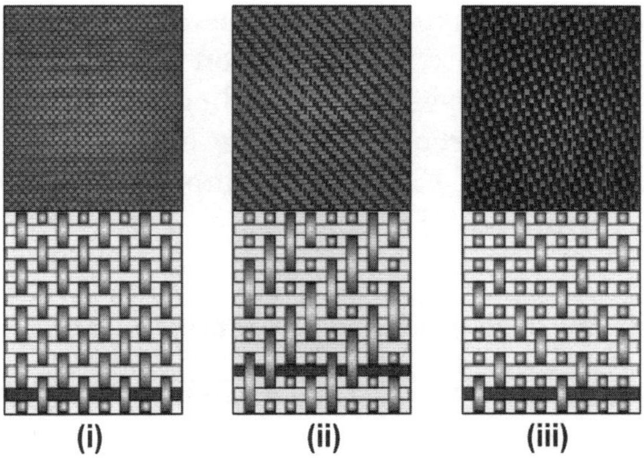

Figure 1.2. *Different weave patterns of carbon fabrics: (i) plain (one weft over one warp), (ii) twill (two warp over two weft) (iii) satin (4 H) (one warp over three weft) [www.fibreglast.com]*

Property	Plain	Twill	Satin
Good stability	good	acceptable	poor
Good drape	poor	good	excellent
Low porosity	acceptable	good	excellent
Smoothness	poor	acceptable	excellent
Balance	good	good	poor
Symmetrical	excellent	acceptable	very poor
Low crimp	poor	acceptable	excellent

Table 1.4. *General properties of some types of fabric weave [www.netcomposites.com]*

1.3.1. *Manufacturing methods to create CFRCs*

There are various methods by which CFRCs are be manufactured, such as compression molding, bladder molding, vacuum-assisted resin transfer molding (VARTM), autoclave molding, etc. Compression molding is the most popular and cost-viable method, with minimum infrastructure being required. There are various techniques for manufacturing CFRCs by compression molding (see section A1.1, Appendix I).

1.3.2. *Performance evaluation of composites*

The CFRCs are characterized mainly by their physical (density, void contents, composition, etc.) and mechanical (strength and modulus [tensile, flexural and compressive], interlaminar shear strength, etc.) properties. These are described in section A2.1, Appendix II.

1.3.3. *Tribological properties*

Generally CFRCs are evaluated for tribo-performance in selected wear modes, such as sliding/adhesive, abrasive, fretting, erosive, etc., depending on the desired end use. Adhesive wear mode is the most frequently reported, since the data lead to the decision to select a CFRC for typical applications such as dry bearings, slides etc. Tribo-studies are conducted to test CFRCs against selected counterfaces, with varying roughnesses and under various operating parameters such as pressure (P)/load, speed (v), temperature, sliding duration, and environment (such as humidity, etc.).

Performance is mainly expressed in terms of three output parameters – the coefficient of friction (μ), specific wear rate (K_o) expressed by equation [1.2], and PV_{limit} values.

$$K_0 = \frac{\Delta W}{\rho L d} m^3 N^{-1} m^{-1} \quad [1.2]$$

where:

– ΔW is the weight loss in kg;

– ρ the density of a wearing material in kg/m^3;

– L the load in N;

– d the sliding distance in meters.

The PV_{limit} values (product of selected pressure, P, and speed, V, values) are useful for estimating the capacity of a material to run without failure under selected P and V conditions. The higher the PV, the value better the performance or utility of a composite under harsh operating conditions.

Unfortunately there are no well-accepted standard methods used by researchers for tribo-performance evaluation. There is miscellany in the selection of test set-ups and operating parameters. Hence an absolute comparison of the performance data provided by various researchers is very difficult. Generally the data from the same group of researchers generated in identical conditions are rated as more reliable for the comparison of such materials.

1.4. Tribo-performance of CFRCs: influential parameters

As already discussed, major influential parameters on the performance (including tribological) of CFRCs include:

– the type of matrix;

– the type of carbon fibers (PAN, pitch, etc.);

– the amount of fabric;

– the weave of the fabric;

– the orientation of fabric with respect to the loading direction;

– the fiber-matrix interface;

– the processing technique and parameters, etc.;

– fillers, solid lubricants, engineered surfaces, etc. (especially in the case of tribo-performance).

In this section, the influence of most of these parameters is discussed based on a few composites developed in the laboratory. The two polymer matrices selected are PEI (ULTEM 1000 from GE Plastics, USA) and PES (Gafone B and Veradel) supplied by Solvay Specialities India Private Limited. These are specialty high-performance polymers that are amorphous and have a common solvent, dichloromethane (CH_2Cl_2).

PAN-based carbon fabric of different weaves (Fibre Glast Developments Corporation, USA) was used to develop the composites using the compression molding method. In general, the impregnation technique was used in the manufacturing of composites, unless otherwise specified.

The details of tribo-testing, including test set-up and conditions, etc., are summarized in detail elsewhere [BIJ 07a, BIJ 07b, BIJ 07c, RAT 06, SHA 11a, SHA 11b, SHA 11c, SHA 11d].

1.4.1. *Influence of the processing technique*

CFRCs based on the PEI matrix using mainly two techniques – impregnation and polymer film followed by compression molding – were developed and evaluated as described in sections 1.4.1.1 and 1.4.1.2. The hand lay-up technique was also used for comparison.

1.4.1.1. *Impregnation technique*

The plies were cut from the carbon fabric roll and the open strands from all four sides were sealed with a PTFE-coated glass-fabric tape to avoid fiber misalignment. These plies were individually immersed in separate containers filled with a viscous solution of PEI in Dichloromethane (DMC) for 12 hours (density 1.32 g/cm^3). The containers were properly sealed to avoid the evaporation of the solvent, which was required for adequate wetting of fiber strands with the PEI solution. The plies were taken out carefully to avoid disturbance in the weave. They were then dried in an oven for two hours at 100°C in a stretched condition.

The prepregs were carefully stacked inside the mold to avoid misalignment. A mold-release agent, PTFE-coated glass fabric was placed on the top and bottom of the stacked prepregs.

The temperature of the mold (380–390°C) was attained within two hours. During the total compression time of 20 minutes, two intermittent breathings (each of two seconds) were applied to expel any adsorbed moisture and residual solvent. The machine was then switched off and the composite was allowed to cool to an ambient temperature, keeping the pressure at 7.35 MPa for four hours. The pressure was then released and the composite was removed from the mold.

1.4.1.2. *Film technique*

The fabric layers were sequenced alternately between the films of a polymer with a selected thickness of 0.0254 mm. These stacked layers were then compression molded under a pressure of 2 MPa at a temperature of 360–370°C.

1.4.1.3. *Hand lay-up technique*

The plies of fabric in stretched condition were painted with the viscous solution of polyetherimide in DCM and were then dried. The strokes were applied in one direction only. Two strokes were generally enough to acquire the desired amount of fabric in the final composite.

Weave/ properties	Plain	Plain	Twill	Twill	Satin	Satin
Processing technique	I	F	I	F	I	F
Composites designation	P_{65}	FP_{63}	T_{65}	FT_{63}	S_{65}	FS_{63}
Density (g/cm^3) ASTM D 792	1.55	1.53	1.53	1.52	1.54	1.54
Contents of fabric, vol./wt %	55/65	52/63	55/65	52/63	55/65	52/63
Tensile strength (MPa) ASTM 638	535	471	**888**	430	575	330
TM (tensile modulus) (GPa) ASTM 638	73	73	**106**	69	76	53
Elongation (%) ASTM 638	0.54	0.58	0.08	0.81	0.32	0.87
Toughness (MPa) ASTM 638	**3.8**	2.2	2.2	2.4	2.8	2.6
flexural strength (MPa) ASTM 790	589	270	**951**	245	832	333
flexural modulus (MPa) ASTM 790	40	42	**54**	29	46	52
ILSS (MPa) ASTM 2344	49	18	**66**	12	63	15

Bold letters show the highest values

Key: F – film technique, I – impregnation technique, ILSS - interlaminar shear strength, P – plain weave, S – satin weave, T – twill weave

Table 1.5. *Performance properties of composites based upon the different processing techniques using the compression molding method [BIJ 06a, RAT 06b]*

1.4.1.4. *Performance evaluation of the CFRCs developed*

Performance evaluation of these composites was carried out as described in Appendix II (section A2.1) and the data are presented in Table 1.5. The data indicate that the impreganation technique led to the creation of composites with superior performance properties.

1.4.2. *Influence of fabric contents*

The Polyetherimide-Carbon fabric (PEI-CF) composites were prepared using the impregnation technique and plain weave of carbon fabric, as already described. The amount of fabric in the composites was manipulated by varying the density of the PEI solution [BIJ 07a, RAT 06].

1.4.2.1. *Performance evaluation of CFRCs developed*

The composites exhibited the properties shown in Table 1.6. Carbon fabric reinforcement has led to a significant improvement in all the properties. The amount of carbon fabric, however, proved to be a crucial factor. When the carbon- fabric content was too high or too low, the effects were less helpful.

Excessive fibers result in inadequate wetting by the matrix which leads to more damage. Carbon fabric with 85 vol. % proved the poorest choice in almost all properties except tensile strength and modulus. If fibers make up too little of the composite, such as 40 vol. % – as in the case of PEI-CF_{40} – not enough strength (tensile strength) can be imparted since the composite has excessive resin.

As seen in Table 1.6, a moderate percentage of carbon fabric appears to produce composites with the greatest performance properties. Most of the properties were best for 65 vol. % of carbon fabric, followed by 75 vol. %, indicating that the 65–75 vol. % range of carbon fabric optimal for the production of composites expressing the best properties.

Fabric content	PEI	PEI-CF_{85}	PEI-CF_{75}	PEI-CF_{65}	PEI-CF_{55}	PEI-CF_{40}	HP_{69}
Density (g/cm^3) ASTM D 792	1.27	1.59	1.58	1.57	1.55	1.49	1.58
Fabric contents (vol./wt. %) ASTM 3171-06	-	85/90	75/80	65/72	55/65	40/50	69/75
Tensile strength (MPa) ASTM 638	105	562	691	**697**	535	330	540
Tensile modulus (GPa) ASTM 638	3.0	76	85	**87**	73	54	62
Toughness (MPa) ASTM 638	-	0.25	1.8	1.4	**3.8**	3.0	0.89
Flexural strength (MPa) ASTM 790	150	367	**1013**	818	589	505	620
Flexural modulus (GPa) ASTM 790	3.3	22	**56**	50	40	29	24
Interlaminar shear strength (MPa) ASTM 2344	-	33	45	**54**	49	35	46

Bold numbers denote the highest quality of a property

Underlined numbers denote the lowest quality of a property

HP_{69} – a composite produced using a hand lay-up technique

CF – carbon fabric. The numbers in subscript represent the percentage of carbon fibers included in the composite

Table 1.6. *Physical and mechanical properties of a series of PEI-CF (plain weave fabric) composites with variation in proportion in of carbon fibers [BIJ 07a, BIJ 07b, BIJ 07c, IND 06, RAT 06]*

CF-PEI_I (composite created by the impregnation technique) and HP_{69} (composite created using the hand lay up technique) contained almost an equal amount of carbon fabric. However, CF-PEI_I exhibited superior performance, indicating that impregnation technique leads to stronger composites compared to the hand lay-up technique.

1.4.2.2. *Tribo-performance*

Tribo-performance was evaluated in various wear modes, as described in [BIJ 07a, BIJ 07b, BIJ 07c, RAT 06, SHA 11a, SHA 11b, SHA 11c, SHA 11d].

1.4.2.2.1. Adhesive wear studies

The carbon fabric enhanced the strength, wear and friction performance of PEI substantially. None of the composites, however, proved to be best in all of the categories relating to strength properties. The composite containing 85 vol. % of carbon fabric proved to perform the worst in almost all the properties, while that containing 65 vol. % proved to be the most effective at enhancing most of the properties, except toughness and elongation to break.

A very high (85 vol. %) or very low (40 vol. %) fabric content proved detrimental in achieving the best combination of strength and tribo-performance.

Moderate fabric contents, such as 55, 65 and 75 vol. % showed significant enhancement and proved promising in achieving the best combination of properties.

The composites did not show significant wear under normal loads and exhibited very high tribo-potential under severe sliding conditions (pressure 4–6 MPa and speed 1 m/s). The wear rates of the three composites (PEI-CF_{55}, PEI-CF_{65} and PEI-CF_{75}) were very low (in the range of 10^{-16} m^3/Nm), which increased slightly with the increase in load along with an appreciably low μ value (0.18–0.30). Taking all the performance parameters into account, carbon fabric content in the range of 65 vol. % was found to be optimum for achieving the combination of very good strength and tribological performance in adhesive wear mode (Figure 1.3).

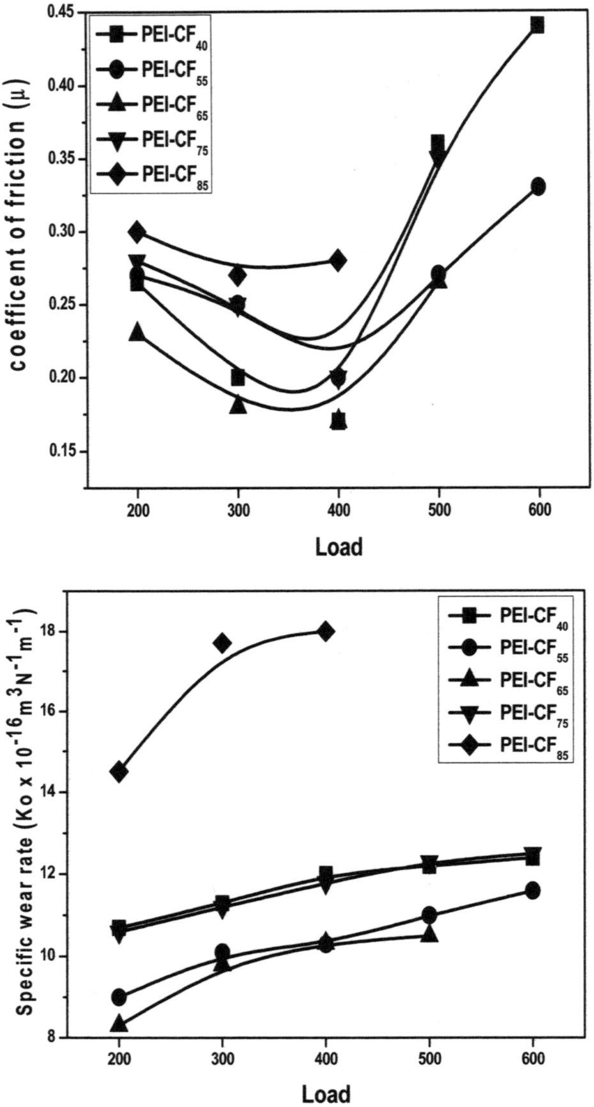

Figure 1.3. *(a) Coefficient of friction; and (b) specific wear rate as a function of load for PEI composites containing varying amounts of carbon fabric [BIJ 07b, BIJ 07c]*

A fairly good correlation was obtained between wear resistance (W_R) and combination of mechanical properties, such as ultimate tensile strength (S), and interlaminar shear strength (ILSS), as shown in Figure 1.4. Scanning electron microscopy (SEM) studies, shown in Figure 1.5, supported the wear behaviour exhibited by the composites. Various wear mechanisms, such as longitudinal wear-thinning of fibers, peeling off, fiber-matrix debonding, etc., were observed to be responsible for the difference in the wear performances of the composites.

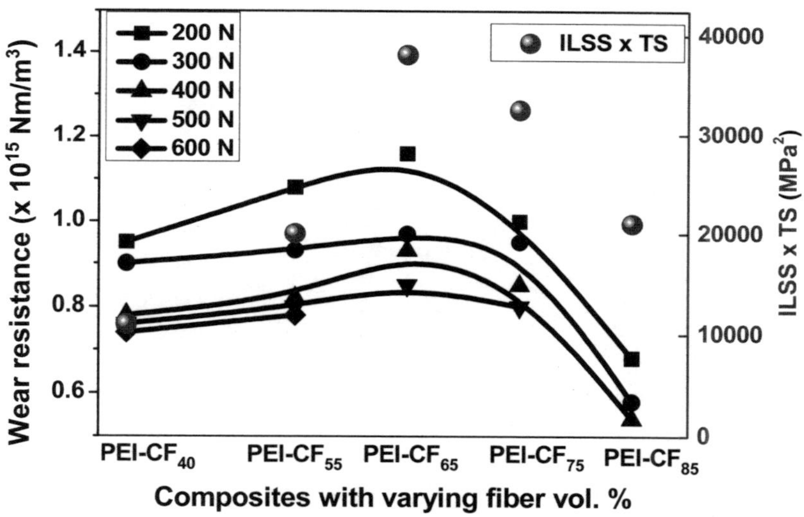

Figure 1.4. *Wear property correlation – wear resistance as a function of the product of tensile strength and interlaminar shear strength (ILSS) of the PEI-CF composites [BIJ 07b]*

The studies on wear mechanisms by SEM supported the difference in wear behavior, as seen in Figure 1.5 [BIJ 07b, BIJ 07c].

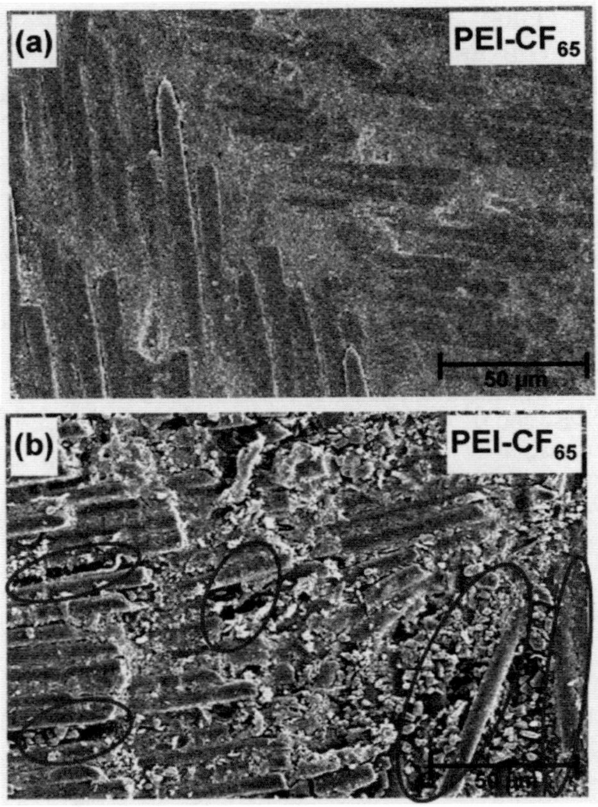

Figure 1.5. *SEM micrographs of worn surfaces of the best and poorest composites: (a) PEI-CF$_{65}$ – weave with warp and weft fibers is visible; very smooth topography; gradual, longitudinal wearing of carbon fabric (fibers originally cylindrical, resulting in flattened surfaces); least de-bonding of fibers and matrix after wearing under thermal and mechanical stresses; no peeling off of fibers supporting the highest W_R of PEI-CF$_{65}$; (b) PEI-CF$_{85}$ – weave with warp and weft fibers is visible; rough topography, so peeling off of fibers; excessive fiber breakage and brittle fracture of matrix; less bonding between the matrix and the fibers supporting lowest W_R [BIJ 07a] (©2007 Elsevier)*

1.4.2.2.2. Abrasive wear studies

In the abrasive wear mode evaluation of composites, specific wear rates (K_o) for these composites were in the range of 2 to 18 × 10^{-10} m^3/Nm. PEI showed the lowest wear rate, indicating that fiber reinforcement deteriorated the W_R of PEI. The difference in W_R between the neat polymer and the composite was attributed to the fact that the abrasive wear of FRPs is dominated by the fiber damage to various extents, depending on the operating conditions, while that of the neat polymer was due to the plastic deformation of a matrix [BIJ 90, JAI 80].

Figure 1.6. *SEM of worn surfaces of composites abraded under 40N, 120 grade (118 µm) SiC abrasive paper; abrading distance 1.5 m: (a) PEI-CF*$_{65}$ *– the fibers are tightly held by the matrix, supporting less wear; micro-ploughing in the composite can be seen in the form of furrows; and (b) PEI-CF*$_{85}$ *– very poor wetting of fibers by the matrix; large number of broken fibers; cross-over point of the plain weave is visible; loosely held fibers without adequate matrix are prone to damage and hence high wear [RAT 06a] (©2006, Springer)*

The most probable mechanisms during abrasion and accounting for wear were: shearing by grits along or across the fibers; fiber-matrix de-bonding; fiber micro-cracking, micro-cutting and micro-pulverization followed by lifting of the fiber debris and pieces from the matrix. The last mechanism leads to random orientations and their subsequent removal from the surface, accounting for wear and leaving behind cavities of appropriate size due to fiber consumption, as can be seen in Figure 1.6.

1.4.2.2.3. Fretting wear studies

The following parameters were selected for the studies:

– load: 100, 150, 200, 250 and 300N;

– stroke length 1 mm;

– oscillating duration two hours;

– oscillating frequency 50 Hz;

– temperature 25°C;

– sliding distance 720 m.

Composites with the highest and lowest amounts of carbon fabric showed poor performance. The composite with 85 wt. % of carbon fabric was the poorest in this wear mode. PEI-CF_{65} proved the best, showing the highest limiting load and tribo-utility in this wear mode. The relative wear resistance (RWR) is a parameter that is very useful for such comparisons. It directly highlights the enhancement in performance (see Table 1.7). RWR is calculated as follows:

$$\text{RWR} = \frac{\text{Wear resistance of composite}}{\text{Wear resistance of matrix}} \qquad [1.3]$$

Material	PEI	PEI-CF_{85}	PEI-CF_{75}	PEI-CF_{65}	PEI-CF_{55}	PEI-CF_{40}	HP_{69}
RWR	1	2.30	3.67	**8.20**	5.51	3.04	**3.7**

Bold numbers denote the highest quality of a property

Underlined numbers denote the lowest quality of a property

HP_{69} – a composite produced using a hand lay-up technique

CF – carbon fabric. The numbers in subscript represent the percentage of carbon fibers included in the composite

Table 1.7. *Relative wear resistance of composites based upon different fabric contents [RAT 06b]*

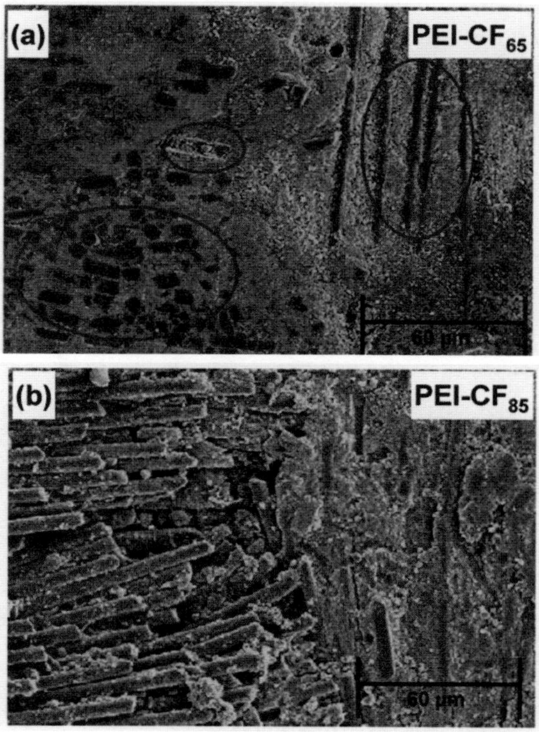

Figure 1.7. *SEM of the worn surfaces of composites fretted under 100 N with a stroke length of 1 mm, oscillating duration of two hours, oscillating frequency of 50 Hz and sliding distance of 720 m: (a) PEI-CF65 composite – gradual and longitudinal wearing of carbon fabric, and cavity filled with wear debris (circled in the middle); and (b) PEI-CF85 composite – edge of a crater showing the brittle fracture of carbon fabric [BIJ 06b] (© 2006, Springer)*

The W_R of the composites increased with the amount of carbon fabric up to a particular value and then started to fall. The optimum amount of carbon fabric for the best wear performance was in the range of 65 wt. %. Interestingly, the ILSS and W_R of composites at 100 and 150 N showed perfect correlation [BIJ 04]. The composite created using the hand lay-up technique (HP_{69}) when compared with PEI-CF_{65} showed a very significant difference in RWR, confirming that the impregnation technique is superior to the hand-lay up technique.

The wear mechanisms in best and poorest composites are shown in Figure 1.7.

1.4.2.2.4. Erosive wear studies

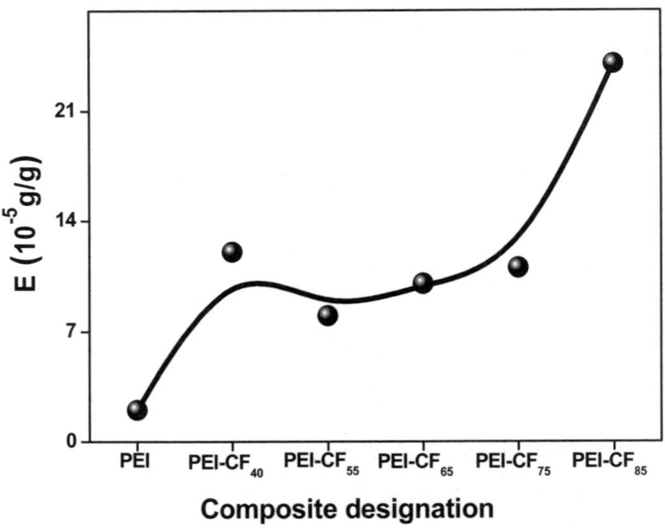

Figure 1.8. *Erosive wear rate of the composites eroded in identical conditions: 480 g of erodent silica sand (106–120 μm grains) with a flux rate of 8 g / min, particle impact velocity of 26.88 m / s and 30° angle of impingement [BIJ 08]*

Tribo-performance evaluation in erosive wear mode (Figure 1.8) indicated a strong dependence on the amount of carbon fabric in a composite. The inclusion of carbon fibers redued the performance of the PEI, as in the case of abrasive wear mode. PEI-CF_{85} proved to be the poorest, followed by PEI-CF_{40}. The other three composites performed far better than these two, confirming that an amount of fabric in the 55–75 vol. % range is the most suitable for good wear performance.

1.4.3. *Fabric orientation effect*

In the case of BD composites, there are two possible orientations of fabric in the composite with respect to the sliding plane. The first is when fabric is parallel to the sliding plane and the second is when fabric is perpendicular to it. In each case, three sub-orientations are possible if the directions of the warp and weft fibers are considered with respect to sliding direction. Thus, a total of six orientations are considered in the case of BD composites [BIJ 02, BIJ 06a]. It has been reported that the orientations of Aramid fabric with respect to abrading direction and plane have played a key role in the case of Aramid-PES composites [BIJ 06a].

In the case of CFRCs, during adhesive wear studies it has been observed that when fabric is normal to the sliding plane (50% of fibers perpendicular to the plane and 50% parallel to the plane), the experiment could not be continued for more than 10 minutes. This is because of the continuous increase in μ, vibrations in machine and an excessive increase in the disc temperature, even at very low load.

When the worn pin surface was inspected, a lot of metallic wear debris of large size was observed to be embedded in the surface, confirming severe abrasion. The carbon fibers used in CFRCs were PAN-based and when the fabric is in the plane parallel to sliding, all fibers – and hence graphene sheets – are parallel to the plane and offer very low resistance to shear, resulting in low μ. When fibers are perpendicular to the sliding plane, however, the graphene planes are also in the direction perpendicular to the sliding plane. Fibers that are oriented in this way show abrasive characteristics instead of lubricating properties.

1.4.4. *Effect of fabric weave on performance properties*

Three CF-PEI composites based on three weaves (plain, twill and satin) and containing an almost equal amount of carbon fibers were developed using two techniques (film and impregnation). The properties of the composites are shown in Table 1.5.

1.4.4.1. *Performance evaluation of the CFRCs developed*

It was interesting to note (Table 1.5) that no weave led to the best properties in all respects. It was also interesting to note that role of weave depended upon the development technique used.

In the case of the impregnation technique, twill weave proved best in 50% of the properties and poorest in the other 50%. Satin weave did not perform the worst in any of the properties, although it came top in only two properties. Plain weave was a moderate performer, with one property being the best and one the worst.

In the case of film technique, plain and satin weaves proved the best for three properties each (tensile strength, TM and ILSS; and toughness, flexural strength and flexural modulus respectively). These weaves were also poorest for two properties (toughness and FM; and tensile strength and TM). Twill weave was a moderate performer and produced the lowest values in flexural strength and ILSS.

1.4.4.2. *Tribo-performance evaluation*

Figures 1.9a and 1.9b show the μ and specific wear rates of composites as a function of load. The μ was in the order

$T_{65} < P_{65} < S_{65}$, indicating that the twill weave is most suited for low composite friction. Interestingly, identical trends were also seen in the case of wear behavior. Twill weave proved best for designing a composite with low μ and low wear rate. This was followed by plain weave. Satin weave proved poorest.

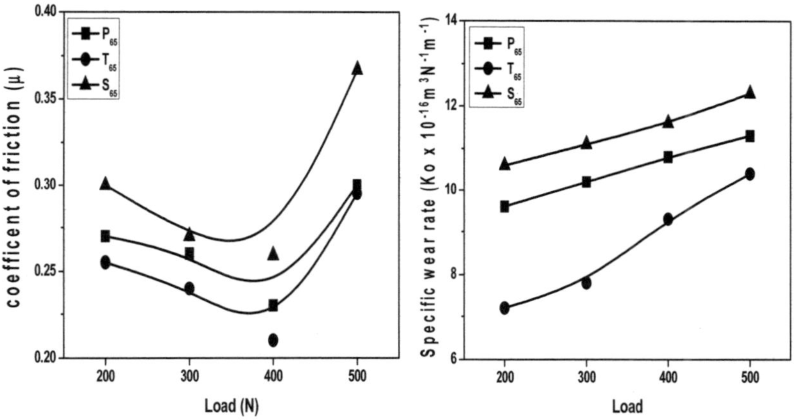

Figure 1.9. *(a) Coefficient of friction; and (b) specific wear rates of PEI composites as a function of load and cabon-fiber weave [BIJ 06b]*

As seen in Figure 1.9b the specific wear rates of the composites were in the range 7 to 12 × 10^{-16} m^3/Nm, which is considered to be very low. It was observed that the weave of the fabric significantly influenced both strength and tribological performance. Twill weave proved to be the best choice for achieving the highest possible strength properties, including tensile modulus, tensile strength, flexural strength, flexural modulus and ILSS (except toughness). It also had the highest wear resistance (W_R) and the lowest μ under various loads. The other two weaves, however, did not show a correlation between strength properties and tribo-performance. It was, thus, concluded that no definite correlation existed between strength and wear properties.

Wear mechanisms, on the other hand, strongly supported the difference in the wear behaviour of composites under various operating conditions [BIJ 06b]. Table 1.8 summarizes the effect of various carbon fabric contents and weaves and that of the processing technique on the mechanical properties of CFRPs.

Composites/ property	Tensile strength	Tensile modulus	Flexural strength	Flexural modulus
variation in contents				
PEI-CF$_{40}$	3.14	18	3.36	18.78
PEI-CF$_{55}$	5.09	24.33	3.12	12.12
PEI-CF$_{65}$	**6.84**	**29**	5.45	15.1
PEI-CF$_{75}$	6.58	28.33	**6.75**	**16.96**
PEI-CF$_{85}$	5.35	25.33	2.45	6.6
variation in weave (impregnation technique)				
P$_{65}$	5.09	24.33	3.92	12.12
T$_{65}$	**8.45**	**35.33**	**6.34**	**16.36**
S$_{65}$	5.47	25.33	5.55	13.93
variation in weave (film technique)				
FP$_{65}$	**4.48**	**24.33**	**1.8**	12.72
FT$_{65}$	4.09	23	1.63	8.78
FS$_{65}$	3.14	17.60	2.22	**15.75**

Bold numbers denote the highest quality of a property

Underlined numbers denote the lowest quality of a property

F – filmtechnique, P – plain weave, S – satin weave, T – twill weave

CF – carbon fabric. The numbers in subscript represent the percentage of carbon fibers included in the composite

Table 1.8. *Relative enhancement factor (REF in percent) for composites based on various carbon fabric contents and weaves developed using different techniques [RAT 06b]*

1.4.5. *Influence of strengthening the fiber matrix interface*

The surface of a carbon fiber is smooth and chemically inert, leading to the problem of inadequate adhesion with the matrix and hence a weaker fiber-matrix interface. Pretreatment of the fiber surface to improve the interface prior to its use in the composite is therefore advantageous [MIC 08, RAN 77, SEL 90]. This leads to an improvement in the properties of composites, including their tribological properties. The surface treatments bring about two major changes on the fiber surfaces.

First, by roughening the surface resulting in a greater surface area and a large number of contact points, micro-pores or surface pits on an already porous fiber surface we get an enhanced fiber-matrix interaction. Interestingly, such treatment adversely affects the strength of fiber. The first effect, considered positive, leads to an enhancement in fiber-matrix adhesion and hence an improvement in the strength of a composite since the matrix supports the fibers more firmly. The other simultaneous effect, which is negative, reduces the strength of the fibers due to etching, and thus contributes to deterioration in the strength of a composite. The final strength of a composite thus depends on the net contribution of these two opposing effects. It is therefore imperative to optimize the extent of treatment to get the maximum possible enhancement in a composite's performance properties.

The second effect, which is purely positive, involves chemical reactions leading to the inclusion of reactive functional groups that promote good chemical bonding with the polymer matrix.

1.4.5.1. *Various treatments selected for carbon fibers*

In this work, the following four surface treatment methods for carbon fibers were selected to examine the comparative effects, and hence on the performance of composites using these treated fabrics.

1.4.5.1.1. Nitric acid treatment

This is the most classical and well proven method for treating carbon fibers, although systematic efforts on dose optimization for achieving the highest possible tribo-performance are unavailable. The carbon fibers are oxidized with concentrated HNO_3, with varying time intervals such as 15, 30, 60, 90, 120, 150 and 180 minutes [TIW 11a]. The resulting fibers were designated F_0, F_{15}, F_{30}, F_{60}, F_{90}, F_{120}, F_{150} and F_{180}.

1.4.5.1.2. Cold remote nitrogen-oxygen plasma

Cold remote nitrogen plasma (CRNP) treatment is a well-known technique for treating carbon fibers [TIW 11b, TIW 11c], although CRNP with traces of oxygen has not yet been exploited in the literature on tribology.

Three treatments created by mixing traces of O_2 (0.5% and 1%) in N_2 – called CRNP, Cold remote nitrogen oxygen or CRNO (0.5%) plasma and CRNO (1%) plasma – were applied to the carbon fibers. The treated fibers were designated F_{P1}, F_{P2} and F_{P3}, respectively.

1.4.5.1.3. Gamma-ray irradiation

The carbon fibers were treated with three doses of γ-ray irradiation measuring 100, 200 and 300 kGy [TIW 11d]. The treated fibers were designated F_{G1}, F_{G2} and F_{G3}, respectively.

1.4.5.1.4. Treatment with nano-particles of rare earth compound (YbF_3)

A novel method based on treating carbon fibers with nano-YbF_3 particles ($\approx$55 nm) in three doses (0.1%, 0.3% and 0.5%) was developed [TIW 11e]. The treated fibers were designated F_{Y1}, F_{Y2} and F_{Y3}, respectively.

1.4.5.1.5. Fiber characterization

Various studies, such as Fourier transform infrared spectroscopy (FTIR), micro Raman spectroscopy (MRS) and X-ray photoelectron spectroscopy (XPS), were conducted on the fibers to evaluate the changes incurred due to the treatments.

FTIR indicated that the various functional groups, such as ether, carboxyl and carbonyl [TIW 11f], were included during these treatments. These groups were responsible for the chemical bonding of PEI or PES with the fibers. Increased fiber-matrix adhesion was also proved based on a simple experiment in which treated and untreated fibers were dipped in a solution of PEI/PES in DMC and taken out before being dried.

The gain in weight due to adherence to the PEI matrix was greater in the treated fibers. The tensile test on fiber tows, however, indicated a weakening of fibers due to the treatment.

MRS studies indicated that the treatment led to an increase in distortion in the graphitic structure of carbon fibers. This promoted the fiber matrix adhesion.

Carbon fibers and other sp^3-bonded amorphous carbons are strong Raman scatterers. The bands with peak position at 1,360 cm^{-1} and 1,593 cm^{-1} are the main features of carbon materials and are called D (disordered) bands and G (graphitic) bands, respectively [MON 02a, NEM 79]. The degree of structural disorder of the carbon fibers' surface was characterized by the ratio of integrated intensity of disorder (I_D) induced to the allowed Raman band (I_G). During treatment of a fiber, the I_D/I_G value increased slightly. Consequently the surface crystallite size (L_a) decreased due to the carbon-fiber treatment [MON 02b]. The graphitic microstructure of carbon fibers is disturbed during surface treatment, the crystallite size is reduced, and the activity of the graphite crystallite boundary is improved [CHA 06].

Figure 1.10 compiles the changes in fibers (MRS and mechanical) due to the optimized doses in each treatment. It can be clearly seen that there was a significant reduction in tensile strength due to all treatments. The nanoparticle treatment proved least damaging to the strength of fibers, followed by plasma, gamma and then nitric acid treatments.

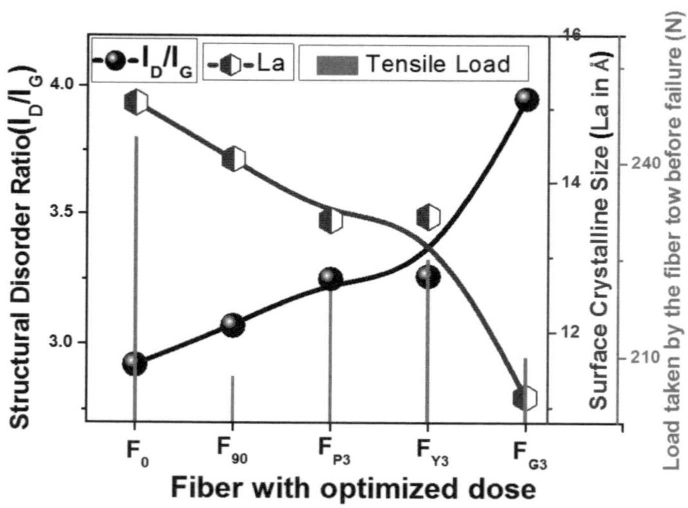

Figure 1.10. *Variation in the structural disorder parameter (I_D/I_G), surface crystalline size (La) and tensile strength of fibers with optimized treatment dose [TIW 11f]*

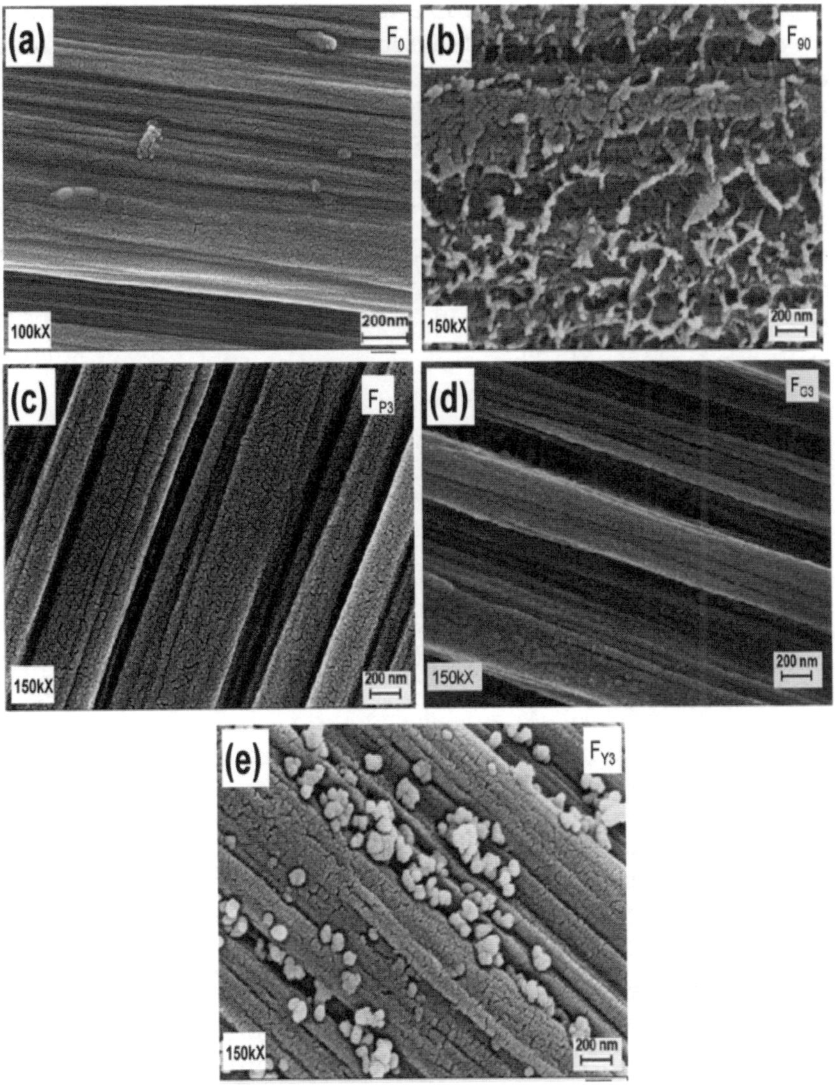

Figure 1.11. *FESEM images of carbon fibers: (a) F_0 – untreated fiber; (b) F_{90} – fiber treated with HNO_3 for 90 minutes; (c) F_{P3} – cold remote nitrogen-oxygen plasma-treated fiber with 1% O_2; (d) F_{G3} – fiber treated with a 300 kGy dose gamma-ray;-and (e) F_{Y3} – fiber treated with 0.3wt % dose of nano YBF_3 [TIW 11b] (©2011 Elsevier)*

The field emission scanning electron microscopy (FESEM) of carbon fibers (one untreated and four with optimized treatment doses) is shown in Figure 1.11.

1.4.5.2. *Composite development*

Several composites based on PEI and (treated and untreated) carbon fabrics were developed. All other parameters such as type and amount of carbon fabric (PAN-based, twill weave, ≈65 wt.%) and the processing technique (impregnation followed by compression molding) were kept constant. This was so that the performance of all of the composites could be compared.

It was observed that the treatments proved successful in improving both mechanical and tribological properties in the case of adhesive and abrasive wear modes. In the case of nitric acid and nanoparticle treatments, typical doses (90 minutes and 0.3 wt %) proved to be optimum and performance declined at higher doses. In contrast to this, in the case of plasma and gamma treatments, the performance increased continuously for all of the selected doses.

1.4.5.3. *Tribological studies*

Figure 1.12 shows the comparative performance of composites based on untreated fabric and that with optimized doses of treated fabric. The figure indicates that the nanoparticle-treated fabric composite proved the best performer, followed by that with treated wtih nitric acid, then gamma radiation and, finally, cold remote nitrogen-oxygen plasma. Although the ILSS of CF_{90} was highest, the wear resistance was not. It was also observed that the difference in W_R was significant due to treatment. The difference in performance due to various treatments, however, was not as significant, although it was substantial for μ and strength properties.

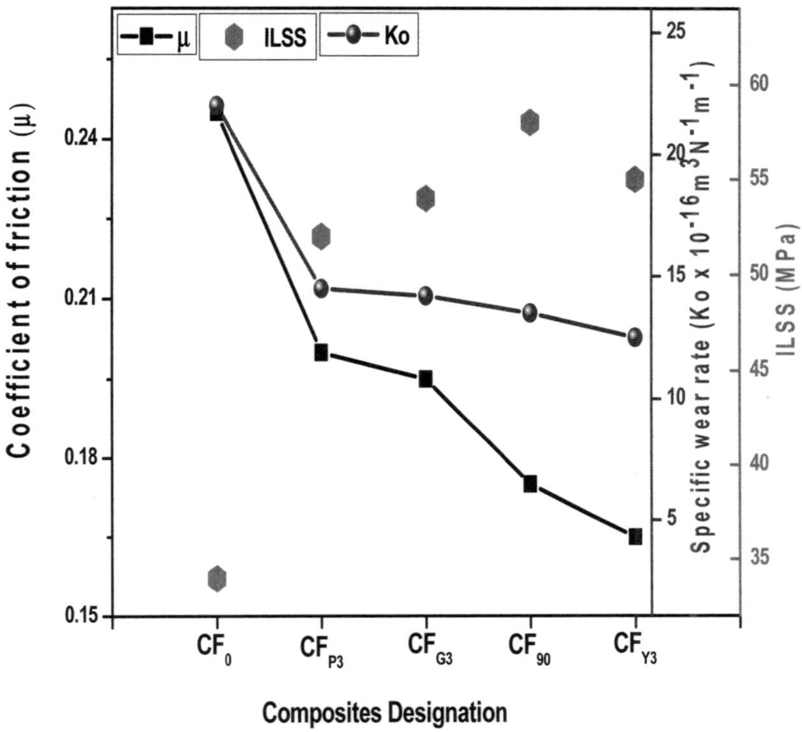

Figure 1.12. *Comparative performance of composites based on untreated fabric and that of composites containing optimized doses of treated fabric at 600 N load and speed 1 m/s [TIW 11a, TIW 11b, TIW 11e]*

Figure 1.13 shows SEMs of worn surfaces, indicating the fibers with an increased amount of resin in micrographs 1.13(b) to 1.13(e) as compared to 1.13(a). The dominant wear mechanisms were fiber fracture, (maximum in 1.13(a)); and wear-thinning of fibers, which leads to enhanced W_R.

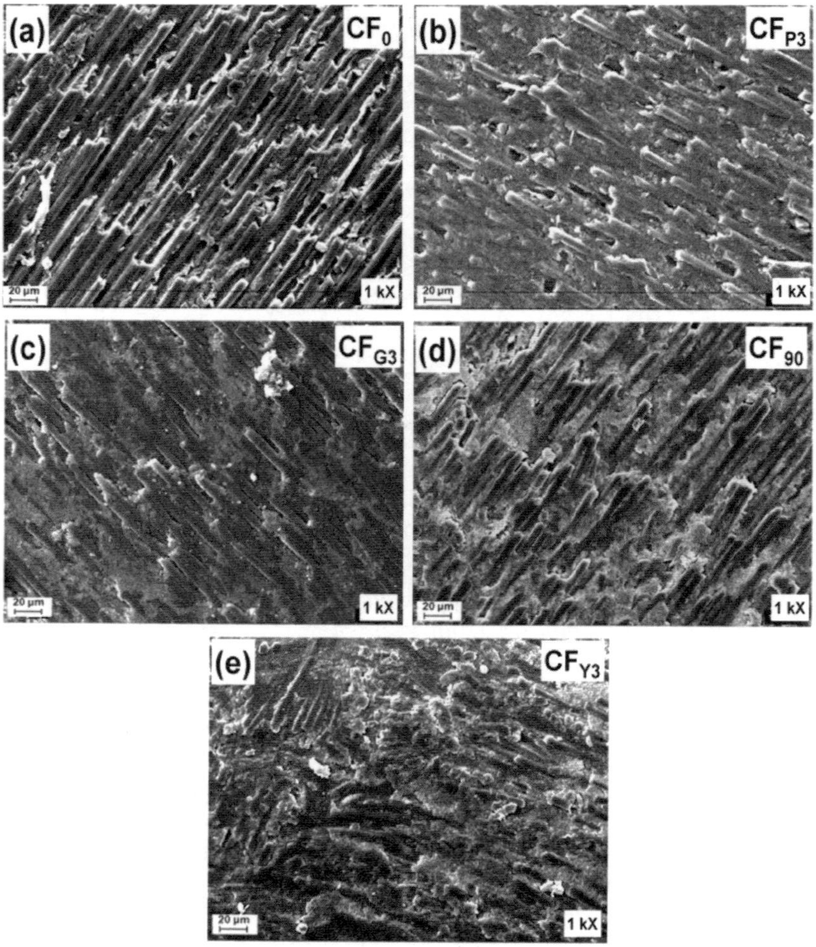

Figure 1.13. *SEMs of composites worn against mild steel under 200 N at 1 m/s. Composites with optimized fiber treatment: (a) CF_0, with untreated carbon fibers; (b) CF_{P3}, with CRNO (1%) plasma-treated carbon fibers; (c) CF_{G3} fibers that γ-ray-treated with 300 kGy; (d) CF_{90} with carbon fibers treated with HNO_3 for 90 minutes; and (e) CF_{Y3}, with carbon fibers treated with nano YbF_3 at a 0.3wt % dose [TIW 11b] (©2011 Elsevier)*

1.4.6. *Influence of the type of polymer used*

The type of a polymer is a major influencing parameter in the performance of CFRCs, especially when all other parameters are constant. Two such composites based on two amorphous thermoplastic polymers – PEI and PES – were developed, keeping all conditions, including the impregnation method (using solutions in DMC), constant. Their performance is described in Table 1.9.

Composite	**PEI**	**PES**	**PEI-CF**	**PES-CF**
Supplier and details of grade of matrices	**GE Plastics USA, ULTEM 1000**	**Solvay Advanced Polymers, India Veradel 3600P**	-	-
Density (g/cm^3) ASTM D 792	1.27	1.37	1.53	1.56
Fabric contents ASTM 3171-06 (wt. %)	-	-	67.15	67.92
Tensile strength (MPa) ASTM 638	**105**	86	750	**783**
Tensile modulus (GPa) ASTM 638	03	**3.22**	67	**78**
Toughness (MPa) ASTM 638	60	60	4.12	**4.9**
Flexural strength (MPa) ASTM 790	**150**	123	**951**	832
Flexural modulus (GPa) ASTM 790	**3.3**	2.9	61	**67**
Interlaminar shear strength (MPa) ASTM 2344	-	-	38	**52**

**Numbers in bold denote the highest quality of a property*

Table 1.9. *Physical and mechanical properties of CRFCs and base matrices*

Barring flexural strength, all properties of PES-based composites were superior to those based on PEI. This indicates that PES is a better matrix as compared to PEI. Interestingly, if the properties of these resins are compared, PEI excels in almost all of the properties measured.

1.4.7. *Influence of the molecular weight of a polymer*

It is important to understand whether the molecular weight of a polymer affects the properties of its composites. PESs with three different molecular weights and properties, including tribological ones, are included in Table 1.10. These PESs were selected to develop three CFRCs, keeping all other parameters the same (percentage of carbon fibers, weave, impregnation technique, etc.). Their performance properties were then characterized, see Table 1.11.

The order for the average molecular weight (MW)and melt flow index (MFI) for pristine polymers was as follows:

MW – $PES_H < PES_M < PES_L$; MFI – $PES_H < PES_M < PES_L$;

With an increase in molecular weight, the melt flow index – and hence flowing tendency of the PES polymer – decreased. Thus, the PES_L has the lowest molecular weight and the highest melt flow index or lowest melt viscosity. The PES_H, however, has the highest molecular weight and the lowest melt flow index or highest melt viscosity.

Overall, the strength and modulus properties followed the trend:

$PES_H > PES_M \geq PES_L$.

Properties/materials	**PES_H**	**PES_M**	**PES_L**
Weight-average molecular weight (M_W)*	46,020	39,825	37,770
Melt flow index (g/10min) ASTM D 1238	33	50	55
Heat distortion temperature (°C) ASTMD 648	204	204	204
Density (g/cm^3) ASTM D792	1.37	1.37	1.37
Tensile strength (MPa) ASTM D638	86	84	82
Tensile modulus (GPa) ASTM D638	3.22	3	2.97
Strain at break (%) ASTM D638	71	60	60
Flexural strength (MPa) ASTM D 790	123	112	112
Flexural modulus (GPa) ASTM D790	2.9	2.8	2.8
Coefficient of friction (at 50 N load, 1 hr sliding time, 1m/s sliding speed)	0.29	0.30	0.33
Specific wear rate (10^{-13} m^3/Nm) (at 50 N load, 1 hr sliding time, 1 m/s sliding speed)	6.3	6.5	8.1

Supplier's M_W data: H – high, M – medium, and L – low

Table 1.10. *Physical and mechanical properties of pristine PES polymers [SHA 11a, SHA 11e]*

Since the difference in the molecular weight of PES_M and PES_L polymers was not significant, the strength performance of the polymers was not significantly different. The wear rate and friction properties followed the same order, confirming that higher the molecular weight, the better the tribo-properties.

Properties/materials	PES-CF_H	PES-CF_M	PES-CF_L
Fiber weight (wt. %) ASTM 3171-06	71.52	68.24	67.92
Density (g/cm^3) ASTM D792	1.58	1.54	1.56
Tensile strength (MPa) ASTM D638	747	778	**783**
Tensile modulus (GPa) ASTM D638	74	76	**78**
Toughness (MPa) ASTM D638	4.4	4.3	**4.9**
Flexural strength (MPa) ASTM D790	736	835	**832**
Flexural modulus (GPa) ASTM D790	63	**68**	67
ILSS (MPa) ASTM D2344	46	46	**52**

M_W: H – high, M – medium, and L – low

Bold numbers represent the highest values

Underlined numbers represent the lowest values

Table 1.11. *Physical and mechanical properties of CF-PES composites reinforced with plasma-treated carbon fibers [SHA 11e]*

As seen in Table 1.11 in the case of CFRCs based on three varieties of PESs, the properties relating to strength were in the opposite order (PES-CF_L > PES-CF_M > PES-CF_H) to that for the pristine polymers ($PES_H > PES_M \geq PES_L$). The reason for this lies in the wetting capacity of the fibers by the matrix, which is a direct function of melt flow index and an inverse function of molecular weight. Figure 1.14 explains the two main events that occur when the viscosity of a solution and the melt flow index affect the wetting of fibers, and hence the performance of composites.

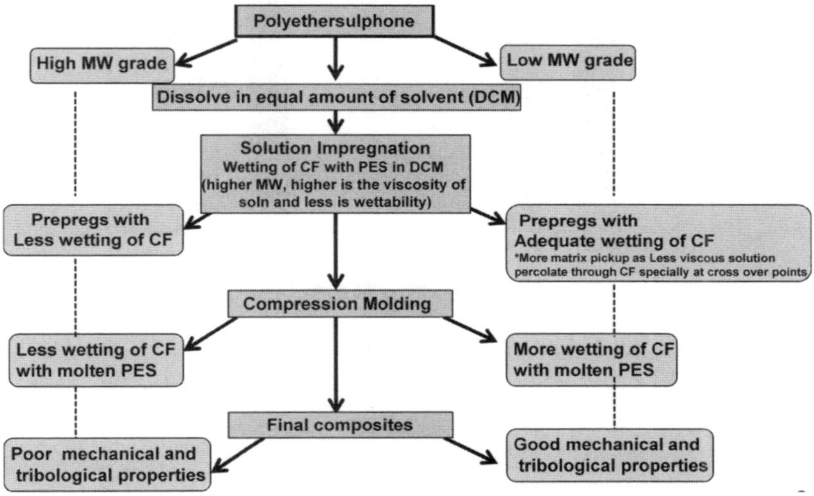

Figure 1.14. *Schematic depicting the effect of viscosity of solution and molecular weight of a polymer on the performance properties of composites during impregnation followed by compression molding (CF – carbon fabrics, DCM – Dichloromethane, MW – molecular weight*

1.4.7.1. *Tribological studies*

Adhesive wear studies against a mild steel counterface under 600 N load and 1 m/s speed indicated that molecular weight had a strong influence on a base polymer in the composite, as seen in Figure 1.15. The higher the molecular weight of a polymer was, the worse the tribo-performance of its composite with CF was. This indicates that the molecular weight of a base polymer plays a vital role in this aspect. A lower molecular weight polymer may be selected for higher tribo-performance if it is used in combination with reinforcement.

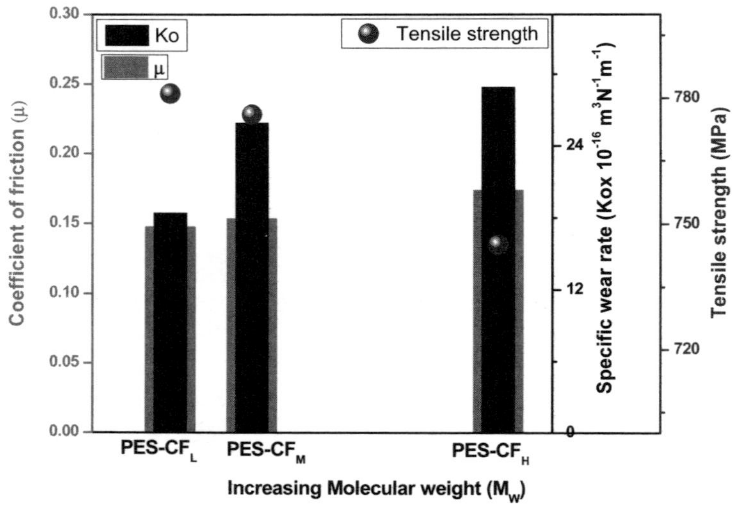

Figure 1.15. *Correlation studies – μ and Ko for PES-CF composites as a function of Mw (at 600 N load and 1 m/s speed against mild steel) [SHA 11e]*

1.5. Concluding remarks

Carbon fiber, especially in the form of fabric, is the most sought-after reinforcement in structural and tribological applications in situations where oils cannot be used as lubricants. The performance properties of carbon CFRCs depend on various parameters, such as:

– the type of matrix and its molecular weight;

– the type of carbon fibers used (PAN, pitch, etc.);

– the amount of fabric;

– its weave;

– its orientation with respect to loading and sliding directions;

– its fiber-matrix interface;

– its processing technique;

– etc.

In-depth studies were taken up as they provided an insight into the extent of influence of such parameters on the physical, mechanical and tribological properties of composites. Various series of composites, by varying one of the parameters and keeping others constant, were developed and evaluated for performance, including tribological in various wear modes.

In the case of adhesive and fretting wear modes, a significant enhancement in wear-performance was observed due to carbon fiber inclusion (by almost an order of three), apart from a reduction in μ and an enhancement in PV_{limit}. In the case of abrasive and erosive wear situations, however, reinforcement proved detrimental. This confirmed the fact that not all types of reinforcements are suitable in all types of wear modes.

From among the three methods for processing, impregnation proved the best in almost all performance properties, including tribological. This was followed by hand lay-up processing, while film technique proved the poorest, since wetting of fibers – especially at cross-over points – was inadequate.

It was observed that in the case of a composite with PAN-based carbon fibers, the orientation of fibers should be parallel to the sliding plane for the best possible tribo-performance.

In the case of selecting the contents of fabric in a composite, in general it was observed that very high or very low fiber contents were undesirable from the points of view of strength and tribo-performance. In general, a moderate amount ≈65% (by vol.) proved to be optimum for various combinations of properties. The overall performance order of composites (fabric percentage wise) was: 65 > 55 > 75 > 40 > 85 (this order was identical in the case of ILSS). Thus, the 65–55 vol. % range proved the most desirable and, depending on the selected wear mode, a slight change in the ranking of these two series of composites was observed (see Table 1.12).

Wear mode	**Adhesive**	**Fretting**	**Abrasive**	**Erosive**
Processing technique	I > F	I > H > F	I > F	I > F
Amount of carbon fibers in composites (vol. %)	65 > 55 > 75 > 40 > 85	65 > 55 > 75 > 40 > 85	65 = 75 > 55 > 40 > 85	55 > 65 > 75 > 40 > 85
Weave	T > P > S	P > T > S	S > T > P	P > S > T
Composites containing fibers with an optimized dose of treatment	$F_{Y3} > F_{90} > F_{G3} > F_{P3}$	-	$F_{Y3} > F_{90} > F_{G3} > F_{P3}$	-
Molecular weight# of polymer	H > M > L	H > M > L	-	-
Composites with carbon fiber and PES of different molecular weights	L > M > H	L > M > H		-
Polymer	PES > PEI	PES > PEI	PES > PEI	-

Processing technique: I – impregnation, H – hand lay-up, and F – film

Weave: T – twill, P – plain, and S – satin

Composites containing fibers with an optimized dose of treatment: F_{90} – fiber treated with HNO_3 for 90 minutes, F_{P3} – fiber treated with cold remote nitrogen-oxygen plasma (1% O_2), F_{G3} – fiber treated with a 300KGy dose of γ-ray , and F_{Y3} – fiber treatd with with a 0.3wt % dose of nano YbF_3 ;

Molecular weight: H – high molecular weight, M – moderate molecular weight, and L – low molecular weight

Table 1.12. *Summary of the parameters influencing the tribo-performance of CFRCs*

In the case of weaves, the wear performance of composites very much depended on the selected wear modes and processing technique used, as seen in Table 1.12. No general recommendation for the selection of weave of fabric was able to be made. For strength properties, twill weave proved superior in most cases, followed by satin weave. Plain weave proved the poorest choice in almost all selected strength properties.

In the case of treatments to strengthen the fiber-matrix interface, while comparing the capability of various optimized doses it was observed that nanoparticle (0.3 wt. % YbF_3)-treated fabric was the most suitable for improving strength and tribological performance (see Table 1.12). This was followed by nitric acid treatment (for 90 minutes). Fabric treated with gamma radiation (300 KGy) proved to be the third choice, followed by cold remote nitrogen-oxygen plasma-treated fabric. The nano-particle treatment and HNO_3 treatment are very cost effective and need hardly any expensive infrastructure, while the other two methods need expensive infrastructure. The radiation treatment requires special precautions to be taken from health and safety point of view. Gamma radiation and plasma treatment did not prove as effective at strengthening the interface of composites.

The molecular weight of a selected polymer proved to be an important parameter. In the case of virgin polymers, a higher molecular weight is advised for superior mechanical and tribological properties, while for carbon fiber-reinforced composites the reverse trends were observed. In three selected grades, the composite containing polymer with the lowest molecular weight as a matrix, showed the best performance.

In the case of two selected amorphous high-performance polymers, PESI proved to be a better choice for developing composites than the PEI.

This chapter thus presents important guidelines (based on research carried out in the authors' laboratory) to design BD composites using carbon fabric and selected polymers as a matrix.

1.6. Bibliography

[BIJ 90] BIJWE J., TEWARI U.S., VASUDEVAN P., "Friction and wear studies of bulk polyetherimide", *Journal of Material Science*, vol. 25, p. 548-556, 1990.

[BIJ 02] BIJWE J., INDUMATHI J., SATAPATHY B.K., GHOSH A.K., "Influence of carbon fabric on fretting wear performance of PEI composites", *ASME Journal of Tribology*, vol. 124, p. 834-840, 2002.

[BIJ 04] BIJWE J., AWTADE S., SATAPATHY B.K., GHOSH A., "Influence of concentration of aramid fabric on abrasive wear performance of polyethersulfone composites", *Tribology Letters*, vol. 17, p. 187-194, 2004.

[BIJ 06A] BIJWE J., AWTADE S., GHOSH. A., "Influence of orientation and volume fraction of aramid fabric on abrasive wear performance of polyethersulfone composites", *Wear*, vol. 260, no. 4-5, p. 401-411, 2006.

[BIJ 06b] BIJWE J., RATTAN R., "Influence of content of carbon fabric on the low amplitude oscillating wear performance of polyetherimide composites", *Tribology Letters*, vol. 23, no. 3, p. 223-229, 2006.

[BIJ 07a] BIJWE J., RATTAN R., "Carbon fabric reinforced polyetherimide composites: optimization of fabric content for best combination of strength and adhesive wear performance", *Wear*, vol. 262, no. 5-6, p. 749-758, 2007.

[BIJ 07b] BIJWE J., RATTAN R., "Influence of weave of carbon fabric in polyetherimide composites in various wear situations", *Wear*, vol. 263, p. 984-991, 2007.

[BIJ 07c] BIJWE J., RATTAN R., FAHIM M., "Abrasive wear performance of carbon fabric reinforced polyetherimide composites: Influence of content and orientation of fabric", *Tribology International*, vol. 40, no. 5, p. 844–854, 2007.

[BIJ 08] BIJWE J., RATTAN R., FAHIM M., TIWARI S., "Erosive wear of carbon fabric reinforced polyetherimide composites: Role of amount of fabric and processing technique", *Polymer Composites*, vol. 29, no. 3, p. 337-344, 2008.

[BUN 88] BUNSELL A.R., *Fibre Reinforcements for Composite Materials*, Elsevier Science Publishers BV, Amsterdam, The Netherlands, 1988.

[BUR 99] BURCHELL T.D., *Carbon Materials for Advanced Technologies*, Elsevier Ltd., Amsterdam, The Netherlands, 1999.

[CHO 93] CHOU T.W., *Materials Science and Technology*, VCH Publishers Inc., New York, USA, 1993.

[CHU 94] CHUNG D.L., *Carbon Fiber Composites*, Butterworth-Heinemann, 1994.

[CHA06] CHAUDHURI S.N., CHAUDHURI R.A., BENNER R.E., PENUGONDA M.S., "Raman spectroscopy for characterization of interfacial debonds between carbon fibers and polymer matrices", *Composite Structructure*, vol. 76, p. 375-387, 2006.

[DOS 87] DOSTAL C.A., *Engineered Materials Handbook – Composites*, ASM International, Ohio, USA, 1987.

[DON 98] DONNET J.B., *Carbon Fibers*, Marcel Dekker, 1998.

[FUS 88] FUSARO R.L., "Evaluation of several polymer materials for use as space lubricants in space", *Tribology Transactions*, vol. 31, p. 174-181, 1988.

[FRI 86] FRIEDRICH K., *Friction and Wear of Polymer Composites*, Composites Materials Series 1, Elsevier, Amsterdam, The Netherlands, 1986.

[FRI 93] FRIEDRICH K., *Advances in Composites Tribology*, Composite Materials Series 8, Elsevier, Amsterdam, The Netherlands, 1993.

[IND 06] INDUMATHI J., Friction and wear studies on polyetherimide and composites, Doctoral Thesis, ITMMEC, IIT Delhi, India, 2000.

[JAI 80] JAIN V.K., BAHADUR S., "Surface topography changes in polymer metal sliding", *ASME Transactions*, vol. 102, p. 520-525, 1980.

[KEL 94] KELLY A., *Concise Encyclopedia of Composite Materials*, Pergamon, 1994.

[MAL 08] MALLICK P.K., *Fiber Reinforced Composites: Material, Manufacturing and Design*, CRC Press, Taylor & Francis Group, New York, USA, 2008.

[MIC 08] MICHAEL Q.T., KINGSLEY K.C.H., GERHARD K., SHAFFER M.S.P., BISMARCK A., "Carbon fibre reinforced poly(vinylidene fluoride): impact of matrix modification on fibre/polymer adhesion", *Composite Science & Technolog*, vol. 68, p. 1766-1776, 2008.

[MON 02a] MONTES-MORÁN M.A., YOUNG R.J., "Raman spectroscopy study of high-modulus carbon fibres: effect of plasma-treatment on the interfacial properties of single-fiber-epoxy composites Part II: Characterisation of the fibre-matrix interface", *Carbon*, vol. 40, p. 857-875, 2002.

[MON 02b] MONTES-MORÁN M.A., YOUNG R.J., "Raman spectroscopy study of HM carbon fibres: effect of plasma treatment on the interfacial properties of single fibre/epoxy composites Part I: Fiber characterization", *Carbon*, vol. 40, p. 845-855, 2002.

[MOR 05] MORGAN P., *Carbon Fibers and Their Composites*, CRC press, Taylor and Francis Group, New York, USA, 2005.

[NEM 79] NEMANICH R.J., SOLIN S.A., "First and second order Raman scattering from finite-size crystals of graphite", *Physical Review B*, vol. 20, p. 392-401, 1979.

[RAN 77] RAND B., ROBINSON R., "Surface characteristics of carbon fibres from PAN", *Carbon*, vol. 15, p. 257-263, 1977.

[RAT 06a] RATTAN R., BIJWE J., "Influence of weave of carbon fabric on abrasive wear performance of polyetherimide composites", *Tribology Letters*, vol. 22, no. 1, p. 105-112, 2006.

[RAT 06b] RATTAN R., Investigations on the mechanical and tribological properties of carbon fabric reinforced polyetherimide composites, Doctoral Thesis, ITMMEC, IIT Delhi, India, 2006.

[SEL 90] SELLITTI C., KOENIG J.L., ISHIDA H., "Surface characterization of graphitized carbon fibers by attenuated total reflection Fourier transform infrared spectroscopy", *Carbon*, vol. 28, no. 1, p. 221-228, 1990.

[SHA 11a] SHARMA M., BIJWE J., MITSCHANG P., "Abrasive wear studies on composites of PEEK and PES with modified surface of carbon fabric", *Tribology International*, vol. 44, no. 2, p. 81-91, 2011.

[SHA 11b] SHARMA M., BIJWE J., SINGH K., "Studies for wear property correlation for carbon fabric – reinforced PES composites", *Tribology Letters*, vol. 43 no. 3, p. 267-273, 2011.

[SHA 11c] SHARMA M., BIJWE J., "Influence of fiber-matrix adhesion and operating parameters on sliding wear performance of carbon fabric polyethersulphone composites", *Wear*, vol. 271, p. 2919-2927, 2011.

[SHA 11d] SHARMA M., Carbon fabric reinforced polymer composites: development, surface designing by micro and nano PTFE and performance evaluation, Doctoral Thesis, ITMMEC, IIT Delhi, India, 2011.

[SHA 11e] SHARMA M., BIJWE J., "Influence of molecular weight on performance properties of polyethersulphone and its composites with carbon fabric", *Wear*, 2011.

[SOU 05] SOUTIS C., "Carbon fiber reinforced plastics in aircraft construction", *Material Science & Engineering A*, vol. 412, p. 171-176, 2005.

[TIW 11a] TIWARI S., BIJWE J., PANIER S., "Tribological studies on polyetherimide composites based on carbon fabric with optimized oxidation treatment", *Wear*, vol. 271, p. 2252-2260, 2011.

[TIW 11b] TIWARI S., BIJWE J., PANIER S., "Adhesive wear performance of polyetherimide composites with plasma treated carbon fabric", *Tribology International*, vol. 44, p. 782-788, 2011.

[TIW 11c] TIWARI S., BIJWE J., PANIER S., "Influence of plasma treatment on carbon fabric for enhancing abrasive wear properties of polyetherimide composites", *Tribology Letters*, vol. 4, p. 153-162, 2011.

[TIW 11d] TIWARI S., BIJWE J., PANIER S., "Polyetherimide composites with gamma irradiated carbon fabric: studies on abrasive wear", *Wear*, vol. 270, no. 9-10, p. 688-694, 2011.

[TIW 11e] TIWARI S., BIJWE J., PANIER S., "Role of nano-YbF_3 treated carbon fabric on improving abrasive wear performance of polyetherimide composites", *Tribology Letters*, vol. 42, no. 3, p. 293-300, 2011.

[TIW 11f] TIWARI S., SHARMA M., PANIER S., MUTEL B., MITSCHANG P., BIJWE J., "Influence of cold remote nitrogen oxygen plasma treatment on carbon fabric and its composites with specialty polymers", *Journal of Materials Science*, vol. 46, p. 964-974, 2011.

A1.1. Appendix I: Various techniques for developing CFRCs by compression molding

There are various techniques that can be used for manufacturing CFRCs by compression molding. These are discussed below.

A1.1.1. *Hand lay-up technique*

This is only applicable when a polymer has an appropriate solvent (with a low boiling point and environmentally friendly) and the fibers should not be soluble in it. Fabric of the required dimensions is painted with a unidirectional application of the desired number of strokes of the viscous solution of the matrix with a brush without disturbing the fabric weave pattern. The prepregs prepared are dried and stacked properly to avoid possible misalignments in the orientation of fiber direction during molding. The amount of resin in the composite is adjusted dependent on the viscosity of the solution and the number of coats applied.

A1.1.2. *Impregnation technique*

This is also only applicable for those polymers that have appropriate solvents (with a low boiling point and that are environmentally friendly). This is one of the most common techniques used for developing CFRPs.

The fabric plies are dipped in the solution of a polymer (for few hours) for adequate wetting of the fiber strands. This technique offers enhanced wettability of fibers because the dilute binder solution seeps below the cross-over points and apart from the fiber strands. It thus ensures the higher strength of composites. The only disadvantage is the vulnerability of the sizing agent, which may be sacrificed in the process.

A1.1.3. *Polymer film technique*

The polymer film technique is cost-effective since no solvents are required, as in case of impregnation. It is the easiest technique, since it does not involve any wet and dry prepregs. There are no chances of voids and misalignments in the stacks, since the fabric and films are arranged alternately. Accurate control of the amount of fabric is possible by placing films of known thicknesses. The only major disadvantage lies in the poor wetting capability of molten resin with fiber strands.

A1.1.4. *Powder prepreg technique*

For some specialty thermoplastic polymers, for example PEEK, it is not possible to use the impregnation solution technique to prepare CFRPs since no non-corrosive solvent is available. The powder prepreg method is therefore employed to develop such composites when very fine powder is available. The pieces of fabric are placed on a bed of polymer powder of known weight. This is followed by the placement of an alternate sequence of fabric and powder. The amount of powder is optimised to get the final fiber weight by percentage in the composite. Table A1.1 summarizes the advantages and limitations of each method.

S. No	Technique	Conditions/ situation for applicability	Advantages	Disadvantages/ limitations
1	Hand lay-up	Polymer must have an appropriate solvent*	Sizing of fibers mostly not removed. Difficulty in handling semi-wet fabrics	Matrix cannot access the areas below the cross-over points; composite has "points of weakness". More expensive than 3 and 4, since solvent cost is added (though possibility of recycling exists). Composite may be 2nd best in strength
2	Impregnation/ prepreg	Polymer must have an appropriate solvent*	Best possible wettability of fibers and matrix; therefore strongest composite. Difficulty in handling fully wet fabrics more than in 1	Sizing agent on the fibers may be washed out in the solution. Expensive compared to 3 and 4 since solvents are expensive and not environmentally friendly
3	Film	Solvent or powder not necessary but polymer should be available in film form	Sizing of fibers is intact. No voids possible since no solvents included.Fiber-to-resin ratio can be accurately adjusted. No problem with fabric handling, as in the case of 1 and 2; hence perfect alignment of fabric and film	Polymer cannot access the fiber stands and areas below cross-over points; hence very poor wettability with fibers. Overall very weak interface and "points of weakness" in the composite

4	Powder technique	Solvent not necessary. (Cheaper and environmentally friendly). Polymer should be available in fine powdery form. Used only when no solvent exists or film not easily obtained	All advantages as in case of 3. Cheapest technique	All disadvantages similar to 3. Moreover, difficulty in sprinkling fine powder uniformly on the fabric. Percolation of molten polymer is poorer and less uniform than in 3. Poorest strength of a composite

*low boiling point and environmentally-friendly solvent;
fibers should not be soluble in the solvent

Table A1.1. *The advantages and limitations of various techniques available for manufacturing CFRPs with compression molding*

A2. Appendix II: Characterization methods for CFRCs

A2.1. *Physical characterization*

Density

The density of CFRPs is determined by using the standard Archimedes principle following ASTM D792, as per the following formula:

$$\rho_c = \rho_w \left[\frac{W_{air}}{(W_{air} - W_{water})}\right]$$

where:

– ρ_c is the density of the composites,

– ρ_w is the density of water,

– W_{air} is the weight of the sample in air,

– W_{water} is the weight of the sample in water.

Compositional characterization

The percentage weight and volume of the matrix and carbon fibers in the CFRPs is determined by ASTM 3171-06 with the use of a thimble containing a small piece of sample and inserted into Soxhlet apparatus. The solvent (which has a low boiling point) is used to extract the matrix. The amount of fabric left in the thimble is dried in an oven and weighed. The percentage of carbon fibers and matrix is calculated as follows:

Reinforcement content, W_w (wt. %):

$$W_W \text{ (wt \%)} = \left(\frac{W_f}{W_i}\right) \times 100$$

where W_i is the initial mass of a specimen in grams and W_f is the final mass after extraction in grams.

Reinforcement content, W_V (vol. %):

$$W_V \text{ (vol. \%)} = W_W \times \left(\frac{\rho_c}{\rho_f}\right)$$

where ρ_f is the density of the reinforcement in g/cm^3 and ρ_c is the density of the specimen in g/cm^3.

Matrix content, W_M:

$$W_m \text{ (vol. \%)} = (100 - W_W)$$

$$W_m \text{ (wt. \%)} = (100 - W_V)$$

Some polymers, such as PEEK, do not have appropriate non-corrosive) solvents. In such cases, the ignition loss method (ASTM 2584-02) is used to calculate the fiber weight fraction. The weighed specimen is placed in a silica crucible and ignited. It is left to burn until only ash and carbon fibers are left. The carbonaceous residue is reduced to an ash by heating in a muffle furnace at 565°C, followed by cooling in desiccators before being weighed. Based on the following formula, the specimen's composition is calculated;

$$\text{Matrix content } (W_m) \text{ wt.\%} = \left[\frac{W_1 - W_2}{W_1}\right] \times 100$$

where W_1 is the weight of the specimen in grams and W_2 is the weight of the residue in grams.

Theoretical void content (vol. %):

This is calculated using ASTM 2734 based on the following equation;

$$\text{Theoretical Void content (vol.\%)} = 1 - \rho_c \left(\frac{W_m}{\rho_m} + \frac{W_f}{\rho_f}\right)$$

where ρ_c, ρ_m and ρ_f are the densities of the (experimental) composite, matrix and fiber (supplier's data), respectively; while W_m and W_f are the weight fractions of the matrix and the fiber.

A2.2. *Mechanical properties*

Tensile properties

Tensile strength and tensile modulus, toughness and percentage strain until failure are calculated as per the ASTM D638 standard. For this test, the strain is measured with extensometer; the cross-head speed during the test was 2 mm/min. The tensile strength is calculated from the following equation:

$$\sigma_t = \left(\frac{P}{bd}\right)$$

where P is the maximum load, d is the depth of the specimen and b is the width of the specimen.

Flexural properties

Flexural strength and modulus are calculated according to ASTM D790. Flexural strength was calculated from following equation and the flexural modulus is calculated from the load-deflection diagram.

$$\sigma_b = \left[\left(\frac{3PL}{2bd2}\right)[1 + 6\left(\frac{D}{L}\right)^2 - 4\left(\frac{d}{L}\right)\left(\frac{D}{L}\right)\right]$$

where P is the load, L is the span length, D is the mid-span deflection, d is the thickness of the specimen and b is the width of the specimen.

ILSS properties

ILSS is evaluated according to ASTM D2344, as per the following formula.

$$\text{ILSS} = 0.75\left(\frac{P}{bd}\right)$$

where P is the maximum load, b is the width and d is the depth of sample.

Chapter 2

Adhesive Wear Characteristics of Natural Fiber-reinforced Composites

In this chapter, glass fiber and palm oil fiber were selected as synthetic and natural reinforcements, respectively, for polyester composites. A hand lay-up technique was adopted in order to fabricate a chopped strand, mat glass fiber-reinforced polyester composite (FRPC). Meanwhile, a closed compressed mold was used to fabricate palm oil fiber-reinforced polyester (PORP) composites. Two PORP composites were developed based on untreated or treated (with 6% NaOH) palm oil fibers. These have been denoted UT-PORP and T-PORP composites, respectively. In addition, a neat polyester material was prepared to investigate the effect of each fiber on its tribo-performance.

The tribological performance of both the natural and synthetic fiber-reinforced polyester composites was mainly evaluated in the adhesive wear mode. The FRPC was tested

---

Chapter written by Belal F. YOUSIF.

in three principal orientations according to the orientations of the chopped strand mat (CSM) glass fibers in the matrix relative to the sliding direction of the counterface, i.e. parallel (P-O), anti-parallel (AP-O) and normal (N-O). The adhesive wear tests were conducted under wet and dry contact conditions using a Block-On-Ring (BOR) machine against smooth stainless steel at various sliding distances, sliding velocities and applied loads.

2.1. Introduction

2.1.1. *Why natural fibers?*

Nowadays, there is a tremendous interest in using polymeric composites for tribological applications, such as bearings, gears and bushes. This is due to their superior properties, i.e. high specific strengths and stiffness combined with relatively low densities. In recent decades, there has been a growing environmental awareness across the world, and this has aroused an interest in research into and the development of biodegradable and high-performance materials.

The introduction of natural reinforcing agents to polymeric composites could reduce their influence on the environment. In the past few years, natural fibers have become an attractive option for the reinforcement of polymeric composites, both economically and ecologically. Moreover, they present many advantages over synthetic fibers, such as being renewable, environmentally friendly, low in cost, flexible with regard to their usage, lightweight, naturally recyclable, and biodegradable. Natural fibers such as jute, linen, banana, bamboo, sugarcane, coir and palm oil have been used to reinforce many types of polymers. In the past decade, numerous studies have strived to evaluate the mechanical properties of polymeric composites based on natural fibers and have concluded that they provide an immense support to a vast range of polymers and can be used as alternatives to synthetic reinforcing agents

2.1.2. *Tribology of polymeric composites based on natural fibers*

The tribological performance of polymer composites reinforced with synthetic fibers such as glass, carbon and Kevlar, have been extensively studied by many tribologists. Recently, there has been growing concern with regard to the increase in the rate of depletion of petroleum resources. However, new environmental regulations have been introduced, forcing material designers to find substitutes for synthetic fibers that should be compatible with the environment. Natural fibers exemplify an environmentally-friendly alternatives to conventional reinforcement fibers [ABD 07, BRA 05, DHA 07, JAC 04, OKU 04, LUZ 07, TOW 08], as they present numerous advantages over synthetics ones. For instance, they are:

– obtained from abundantly available renewable resources;

– non-toxic;

– bio-degradable;

– low in cost;

– flexible in usage;

– high in specific strength;

– low in density.

These advantages and the current environmental issues render natural fibers attractive as reinforcement materials for polymers. High-performance polymeric composites containing free fiber resources have thus become an aim of engineering designers for industrial applications.

Natural fiber-reinforced polymer composites are found in many products, such as housing construction materials, furniture and automotive parts. The introduction of natural fibers to tribo-polymeric composites as reinforcement has not been comprehensively considered. In other words, there is a lack of understanding concerning the impacts of natural fibers on the tribological performance of polymeric composites.

The effects of jute, linen, sugarcane and cotton fibers on polymeric composites have been investigated [CHA 06, HAS 07]. Besides this, a few attempts have been made to study the adhesive and abrasive wear behavior of bamboo, since it is a natural composite made up of parenchyma cell-reinforced tissue cells [CHA 07, OKU 04, TON 95, TON 05, TON 98].

For natural (bamboo) composites, several investigations have been conducted into adhesive and abrasive wear modes [CHA 07, OKU 04, TON 95, TON 98, TON 05]. In abrasion mode, the effect of fiber orientation on the 3B-A wear behavior of bamboo has been investigated with grit of varying sizes [CHA 07, TON 95]. In all reported work, the wear performance and friction coefficient of the composites were controlled by the fiber orientations, grit size and applied load.

Furthermore, the generated debris was of the fibrous type when the fibers were parallel to the sliding counterface. The debris was generated after the plowing and micro-cutting of vascular fibers, fiber and matrix fragmentation, fiber delamination, tilting and micro-cutting. Meanwhile, the debris was of particulate type when the fibers were perpendicular to the counterface, i.e. matrix fragmentation was followed by fiber micro-cutting.

The effect of the fiber content on the abrasive wear properties of the cross-section of a bamboo stem has been examined by [TON 05]. The abrasive wear resistance of the stem cross-section was highly dependent on the vascular fiber content of the bamboo, the abrasive particle size and the relative sliding velocity.

In other words, a higher wear resistance of bamboo was found in samples with higher vascular fiber contents. Increased sliding velocity and larger abrasive particles in the bamboo demonstrated a higher wear volume. Furthermore, it was discovered that there was a good relation between certain mechanical properties, such as tensile strength, impact strength, and abrasive wear resistance. A higher mechanical property resulted in a high abrasive wear resistance. In adhesive wear mode, however, only one study (by [TON 98]), has reported on the natural bamboo composites under dry sliding against a gray iron using a BOR machine. Here, the wear volume of bamboo was highly sensitive to normal load, the sliding velocity and the relative orientation of the fibers. The wear volume increased with the normal load and the sliding velocity, and the best orientation was found when the fibers were perpendicular to the counterface. Adhesion, micro-cracking, micro-plowing and micro-cutting were observed on the worn surfaces.

A study of the effect of natural fibers on tribo-polymeric composites was carried out by [ELS 95], who explored linen and jute fiber-reinforced polyester composites in dry adhesive wear mode at different volume fractions of fibers using a pin on disk (POD) machine. In this work, the volume fraction of the fibers was found to control the wear and friction properties. Moreover, a 33 vol.% fraction of fiber enhanced the wear properties and worsened the friction. During the wear process, stray fibers were bent and directed to the sliding direction, without being pulled out of the matrix. This assisted in protecting the polyester region, thus leading to a reduction in material removal.

In another attempt, the effect of jute fibers on the abrasive wear behavior of polyester composites was studied.

The influence of the addition of a maleic anhydride-grafted polypropylene coupling agent on the abrasive wear performance of the composite in two body abrasive wear modes against a 400-grade abrasive paper in multi-pass condition was determined by [CHA 06].

The results demonstrated that the use of a coupling agent gave rise to an improved wear resistance as compared to other cases. The formation of linkages at the interface between the matrix and the jute fibers during deformation played a significant role in the wear process. In other words, interfacial adhesion between the fibers and the matrix controlled the wear performance of jute-polyester composites.

In a recent study by [HAS 07], the adhesive wear performance of graphite-modified polyester–cotton composites using Pin-on-Disc against steel under dry contact conditions was reported. Incorporating cotton fibers in the polyester resin improved the structural integrity of the material, and the addition of graphite in the cotton–polyester composites further enhanced the capability of the material to withstand sliding wear. Cotton fibers reduced the specific wear rate of the polyester, but it also increased the friction coefficient. The latter was significantly reduced with the addition of graphite in the cotton–polyester composite.

In another study by the same authors [HAS 06], ultra-high molecular weight polyethylene (UHMWPE) modified polyester–cotton composites were investigated with respect to dry adhesive wear at various UHMWPE concentrations. The specific wear rate of the polyester composite decreased when it was reinforced with cotton and UHMWPE. Meanwhile, the friction coefficient increased with the addition of cotton and was significantly reduced with the inclusion of a combination of UHMWPE and cotton. A cotton content of 7.41 vol. % was found to be the optimum volume fraction at which the composites exhibited a better wear performance compared to the others. A 14.19 vol. % UHMWPE in a polyester resin reduced the friction coefficient to nearly half of that in the neat polyester, and to approximately one third of that found in a cotton–polyester composite.

2.2. Preparation of polyester composites

Unsaturated polyester resin was used in the current work. This orthophalic unsaturated polyester (Revesol P9509) was pre-promoted for use in an ambient temperature and cured by adding methylethylketone peroxide as a catalyst. Mainly, two polyester composites were fabricated based on natural and synthetic fibers. The natural fibers were obtained from the (deposit) palm oil bunch fibers available in Johor state, Malaysia, while the synthetic ones were glass fibers in the form of mats. A closed mold fabrication method was adopted to develop palm oil fiber-reinforced polyester (PORP) composites. Meanwhile, the hand lay-up technique was used to develop CSM FRPC. For the preparation of the neat polyester, an open mold technique was used (120 mm × 120 mm × 20 mm). Before the fabrication processes, the entire mold surface was coated with a thin layer of wax as the release agent. Subsequently, the unsaturated polyester, mixed with 1.5% hardener, was poured into the mold and cured for 24 hours. For both the FRPC and PORP composites, the volume fraction of fibers in the matrix was similar (45–48 vol. %).

2.2.1. *Preparation of FRPC*

FRPC denotes a synthetic composite of polyester reinforced with CSM glass fibers. CSM is a sheet of randomly dispersed chopped glass fibers held together with a resinous binder, as shown in Figure 2.1. The selected CSM glass fibers had a fiber length of 20–30 mm and a density of 450g/m^2.

Figure 2.1. *Scanning electron micrograph of chopped strand mat glass fibers [ELT 06]*

A hand lay-up technique was adopted to fabricate the FRPCs. In the preparation process, a smooth metal mold, 120 mm × 120 mm × 25 mm in size, was placed horizontally and coated with a thin layer of wax, which was used as a release agent. A paint roller soaked with polyester resin (mixed with 1.5% hardener) was rolled over the mold surface to create the first layer of polyester resin. Subsequently a sheet of CSM was placed over the layer of polyester resin. During the build-up process, entrapped air between the layers was squeezed out using a smooth steel roller, which also ensured that the polyester resin layers were distributed uniformly over the surfaces. Another layer of polyester resin was applied over the sheet of CSM.

By repeating these steps, the FRPC was built up to a thickness of 25 mm consisting of 22 layers of CSM and 24 layers of polyester. The composite block was cured at room temperature for 24 hours. The thickness of the polyester interlayers was about 66.25 μm.

2.2.2. *Preparation of palm oil fibers and PORP composites*

A bunch of palm oil fruit was collected from a farm in the state of Johor. In the preparation of the palm oil fibers, the fruit was crushed into small pieces and then dried under the sun for 24 hours. The dried pieces were washed out and dried again for two days. Subsequently, the fibers were extracted from the clean dried pieces of fruit and cut into lengths of 1.0–1.5 cm. The final step consisted of washing the fibers and drying them in an oven for five hours at 45°C. The diameter of the fiber was about 350 μm, and the cross-section was a single fiber thick.

2.2.3. *NaOH treatment*

The prepared palm oil fibers were soaked in a 6% NaOH solution, mixed with tap water at a temperature of 26 ±2°C for 48 hours. The fibers were rinsed and left to dry at room temperature before being put in an oven for five hours at 45°C. Figure 2.2 shows a scanning electron micrograph (SEM) of the treated palm oil fiber. Compared to the untreated fibers (Figure 2.2a), it can be seen that the outer surface (skin) of the treated fiber has been washed out and the bundle of vascular fibers are clearly visible. This could play an important role in enhancing the interfacial adhesion characteristics of the palm oil fibers.

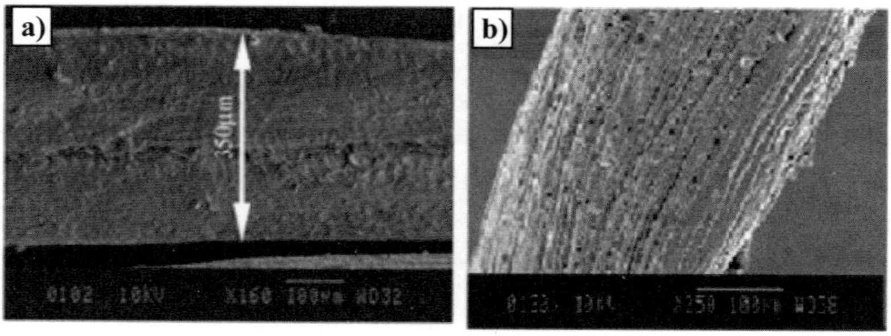

Figure 2.2. *SEMs of a single palm oil fiber: (a) untreated; (b) treated with 6% NaOH [YOU 10]*

2.2.4. *Preparation of PORP composites*

A closed metal mold, 120 × 120 × 20 mm^3 in size, was used to fabricate the PORP composites. The internal surfaces of the mold were greased with a thin layer of wax, which was used as a release agent. The prepared treated or untreated fibers were randomly placed in the mold and pressed. The polyester resin was then poured into the mold until the fibers were totally impregnated. A pressure of about 0.5 MPa was applied to the mold to force out any air bubbles, and the prepared composite block was cured for 24 hours at room temperature (24°C). Two PORP composites, with either untreated or treated palm oil fibers, were fabricated using the above-described method, thus giving UT-PORP and T-PORP.

2.3. Specifications of the fibers and composites

2.3.1. *Interfacial adhesion of palm oil fibers*

Single fiber pull-out tests were conducted on a universal test machine (100Q Standalone) to determine the interfacial adhesion characteristics of untreated and treated fibers. The pull-out test specimens, shown schematically in Figure 2.3, were prepared using a metal mold (50 mm × 10 mm × 10 mm). One end of the fiber was embedded onto the middle plane of a rubber piece that was placed at the end of the mold. The polyester resin was then poured into the mold, and the specimens were cured at 24°C for 24 hours. The length of the fiber embedded in the matrix was about 20 mm; while its free end was about 30 mm. The interfacial shear strength (τIF) was determined by using equation [2.1] [HAR 03, HAW 99, KEN 97]:

$$\tau_{IF} = \frac{F_{\max}}{d_f \cdot \pi \cdot L_e} \quad [2.1]$$

where:

– τ_{IF} = interfacial shear strength,

– $F_{\max}$ = maximum force from force – displacement curve,

– d_f = diameter of fiber; and

– L_e = embedded length.

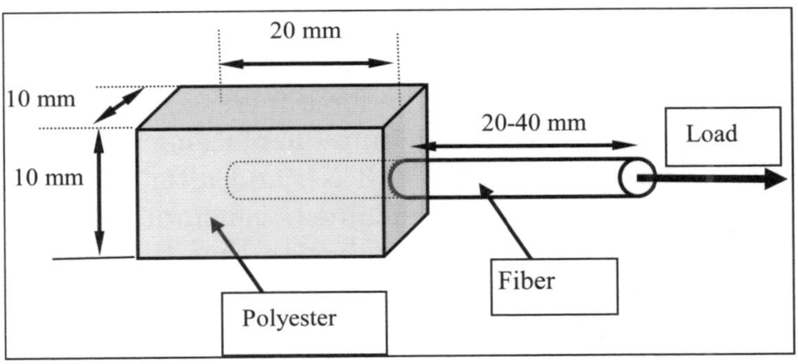

Figure 2.3. *A schematic drawing describing the single pull-out test*

The results of the pull-out tests revealed that the interfacial adhesion characteristic of palm oil fibers was significantly improved when the fibers were treated, i.e. no fiber pull-out was observed during the tests. In other words, the breakage occurred within the fiber rather than the fiber being pulled out of the matrix. Figure 2.4 shows the SEMs of a single treated palm oil fiber after the test. It can be seen that the fiber is still embedded in the matrix, thus indicating that its interfacial adhesion property was good enough to prevent it being pulled out.

Two reasons can explain the higher interfacial adhesion property of the treated palm oil fibers. First, the outer surface (skin) of the fibers was washed out, providing better exposure of the vascular fibers (in the bundle) to the polyester matrix, thereby giving rise to a proper interaction between the fibers and the matrix surfaces. Second, treating the fibers could allow the polyester, in liquid form (during the curing time), to penetrate the bundles through the fine holes in the treated fibers during the preparation of composites. To confirm the second point, a T-PORP composite pin ($10 \times 10 \times 20$ mm^3) was polished at a sliding velocity of 5.7 m/s with a 50 N applied load against smooth stainless steel.

A typical force-displacement curve is shown in Figure 2.5. At the initial loading stage, the force increased linearly with the displacement. At a 1.5 mm displacement, the force reached a maximum value (17.3 N), and after this point it dropped drastically and remained constant, while the displacement increased. This indicates that the pulling-out process was initiated at a 1.5 mm displacement and 17.3 N. The interfacial shear strength (τIF) was found to be about 78 MPa.

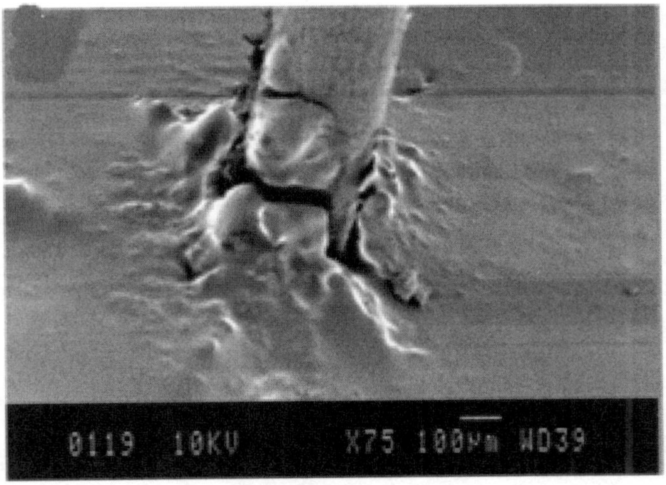

Figure 2.4. *SEM of treated palm oil fibers pulled out of the polyester matrix*

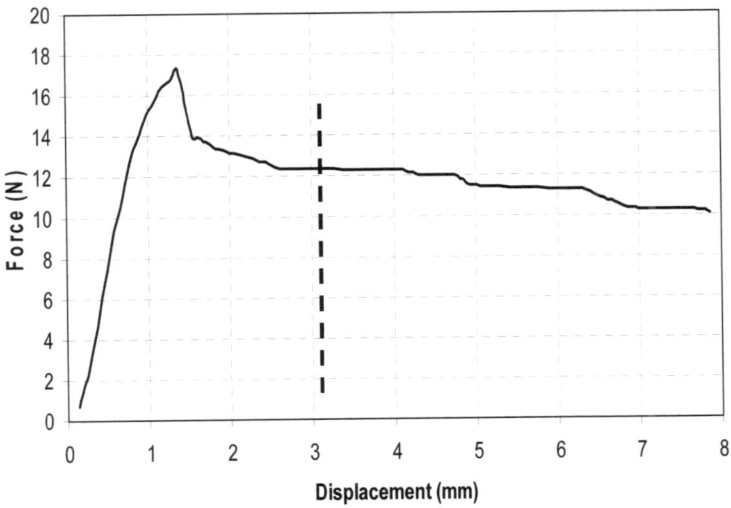

Figure 2.5. *A force-displacement curve of a typical palm oil fiber*

SEMs of the pull-out tests carried out on a sample of untreated palm oil fiber are shown in Figure 2.6. It can be seen that the palm oil fiber was pulled out of the matrix. Moreover, the outer surface of the fiber adhered to the matrix. In addition to this, it could be suggested that the untreated palm oil fibers in the composites did not contain any polyester.

Another explanation for the poor interfacial adhesion of the untreated palm oil fibers is that their outer surface seemed to prevent the polyester form entering the bundles. Thus, the core of the bundles was insulated from the matrix.

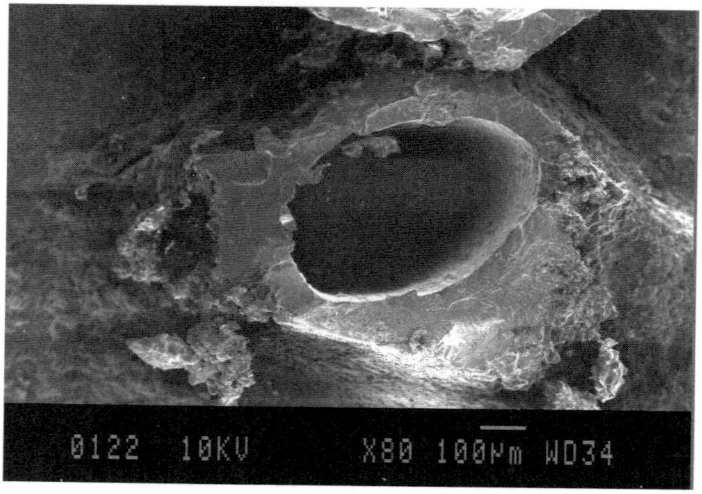

Figure 2.6. *SEM showing untreated palm oil fibers pulled out of the polyester matrix*

2.3.2. *Mechanical properties of the composites*

A universal test system (100Q Standalone) was used to determine the tensile properties of single fibers with a loading speed of 1 mm/min. The tensile properties obtained from glass, untreated and treated palm oil fibers are listed in Table 2.1.

Type of fiber	Tensile strength (MPa)	Initial modulus (GPa)	Elongation at the break (%)
Glass	440	40–45	4.2–5.3
Untreated palm oil	50–55	0.57–0.59	16.2
Treated palm oil	75–80	1.7–1.4	14.2

Table 2.1. *Tensile properties of the various fibers*

The tensile and compression strength of the selected composites were determined with a WP300 PC Aided Universal Material Test machine equipped with computerized data acquisition systems. The Shore D hardness was measured using a TH210 (Time Group, Inc). In order to prevent fracture due to excessively strong gripping forces during tensile experiments, the grip forces were investigated using preliminary tests on a few specimens. The tests were performed at room temperature. For each type of composite, the measurements were repeated three times, and an average was determined. The applied load and displacement were measured simultaneously. The subjected load was incremented at a constant rate of 2 mm/min, and the resulting values of tensile, compression and hardness properties of the neat polyester, FRPC, and the T-PORP as well as the UT-PORP composites are listed in Table 2.2.

Material	σ u (MPa)	E (GPa)	ϵ (%)	Hardness Shore D
Neat Polyester	25	1.3	2.5	6.8
FRPC	85	1.6	6.3	9.62
UT-PORP	47	1.7	4.3	6.9
T-PORP	65	1.65	3.2	7.52

Table 2.2. *Various mechanical results of FRPC, UT-PORP and T-PORP composites, as well as neat polyester*

2.4. Tribo-experimental details

A new tribo-test-machine combining the Block on Disk (BOD) and Block on Ring (BOR) techniques was developed. The main components of the machine were the counterface, two load levers, load cells, the motor, a lubricant container and dead weights. The load cells (Accutec B6N-50 and Accutec H3-50) were fixed on the BOD and BOR load levers to measure the frictional forces between the specimens and the counterface. They were integrated with a weight indicator in order to simultaneously read the friction forces. Moreover, infrared thermometers (Extech model 42580) were attached to the frame of the rig and directed towards the rubbing area in order to measure the interface temperature. For wet tests, the container was filled with tap water in which both specimens were fully immersed. Figure 2.7 shows the developed tribology machine working under dry and wet contact conditions.

Part	Specification
Wheel speed	5 to 2,000 rpm
Wheel type	6.65" steel wheel, as per ASTM B 611
	9" rubber wheel, as per ASTM G 65
Load	5 to 500 N
Motor type	AC motor with frequency inverter drive
Specimen size	25 mm × 58 mm for BOR 10 mm × 10 mm for BOD
Specimen thickness	6 mm to 20 mm
Test types	ASTM G 65, G 105, B 611, G 137-95 and G 77

Table 2.3. *Technical specifications of the newly-developed machine*

Figure 2.7. *The tribo-test machine running in adhesive mode: (a) under dry contact conditions; and (b) under wet contact conditions*

2.4.1. *Experimental procedure*

A series of tribo-experiments was conducted to investigate the adhesive performance of the T-PORP, UT-PORP, FRPC and neat polyester composites. The adhesive tests were carried out using a BOR under dry and wet contact conditions. The FRPC was tested in three principle orientations with respect to the sliding direction, i.e. parallel (P-O), anti-parallel (AP-O) and normal (N-O), see Figure 2.8.

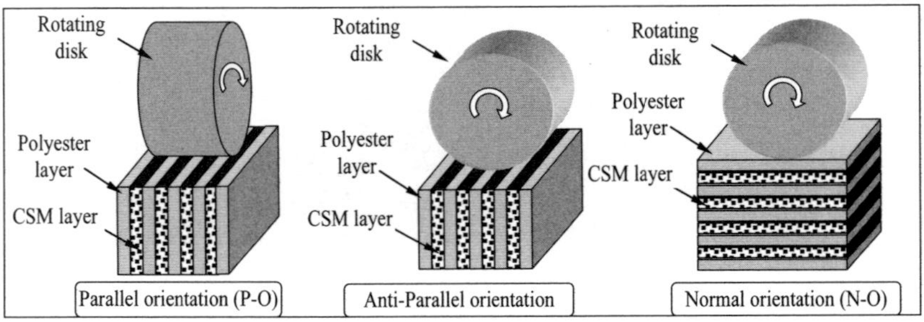

Figure 2.8. *Schematic illustrations of the FRPC specimens displaying the orientations with respect to sliding direction*

Adhesive wear tests of the composites were conducted against a stainless steel disk (AISI 304). Before each test, the disc was polished with abrasive SiC paper (G1500) to a surface roughness of 0.06–0.12 μm Ra. At the same time, the composite specimens were rubbed over an abrasive SiC paper (G1500) to ensure an intimate contact between the sliding face of the specimen and the stainless steel counterface. Sliding tests were conducted at ambient temperature and humidity conditions with varying normal loads (30–100 N), sliding speeds (1.1–7.8 m/s), and sliding distances (0–28 km) under wet and dry contact conditions. Two techniques were considered in the tests: BOD and BOR. In the BOR, the contact was of line type, and it increased as the wear progressed.

For wet tests, the container was filled with tap water (with a hardness of 120–130 mg/l). For each test, new water was used to ensure that the tests were conducted without any residual debris. During the tests, both specimens were totally immersed in the water and the container was covered to prevent splashing and to keep the water level above the specimens. The experiments were conducted for 10–60 minutes at different rotational speeds (500 and 700 rpm, equivalent to 5.6 and 7.8 m/s for BOR). Before and after each test, the specimens were dried in an oven for 24 hours at 40°C and the weight loss was determined.

The weight loss of the composite specimens was determined by weighing the specimen before and after each test using a ±0.1 mg balance (Shimadzu AW120). Friction forces and interface temperatures were recorded every 10 seconds. Calibrations were carried out on both the load cells and the infrared thermometer to ensure accurate results. The weight indicator showed the friction force in mass unit with an accuracy of ±1 mg. For the interface temperature calibration, in its stationary state the disc was heated up and the thermocouple was placed in the rubbing zone. The infrared beam was focused on the interface area and the temperature values measured by the thermocouple and infrared thermometer were plotted, thus giving rise to the calibration equation.

2.4.2. *Examination of worn surfaces*

SEM (JEOL, JSM 840) was used to observe the worn surfaces of the specimens after each test. Before taking the SEMs, the worn surface was coated with a thin layer of gold through ion sputtering (JEOL model, JFC-1600).

2.4.3. *Parameters measured*

The weight losses were determined using a ±0.1 mg balance (Shimadzu AW120) and specific wear rates (m^3/N.m) were calculated using:

$$Ws = \frac{\Delta V}{L.S_D} \qquad [2.2]$$

where:

– Ws: specific wear rate (m^3/N.m);

– V: volume loss (m^3);

– L: applied load (N);

– SD: Sliding distance (m).

2.4.4. *Results and discussion*

Recently it has been reported that the apparent sliding contact area has a significant influence on the wear behavior of composites [ZHA 07]. Such behavior is discussed concerning the BOR technique for which the apparent sliding contact area starts as a line and increases as the wear progresses. The specific wear rate (Ws) and friction coefficient results of T-PORP, UT-PORP, FRPC (in three orientations), as well as neat polyester, at various operating parameters and contact conditions are presented in Figures 2.9 and 2.10.

2.4.5. *Effect of operating parameters*

The Ws of the UT-PORP composite, FRPC (in three orientations: P-O, AP-O and N-O) and neat polyester at sliding velocities of 2.8 and 3.9 m/s for 30 minute equivalents of 5 km and 7 km sliding distances is displayed as a function of the applied load in Figure 2.9. In general, at low applied loads, there is a gradual increase in the Ws for UT-PORP, FRPC (in P-O and AP-O) and neat polyester. However, at higher loads there is no remarkable difference in the Ws. Furthermore, no pronounced effect of sliding velocity on Ws was noticed for any of the composites. The FRPC composite in N-O illustrates a unique trend, i.e. the higher the load, the higher the Ws.

The friction coefficient and the interface temperature of all the materials at various applied loads are presented in Figure 2.10. Here we can see that the highest friction coefficient and interface temperature are obtained for the UT-PORP composite, whereas the other materials displayed relatively lower friction coefficients and interface temperatures. This could be due to the high interaction between the asperities in contact in the case of UT-PORP composite. This theory was supported by the SEM we took of the worn surface, which will be demonstrated in Figure 2.11.

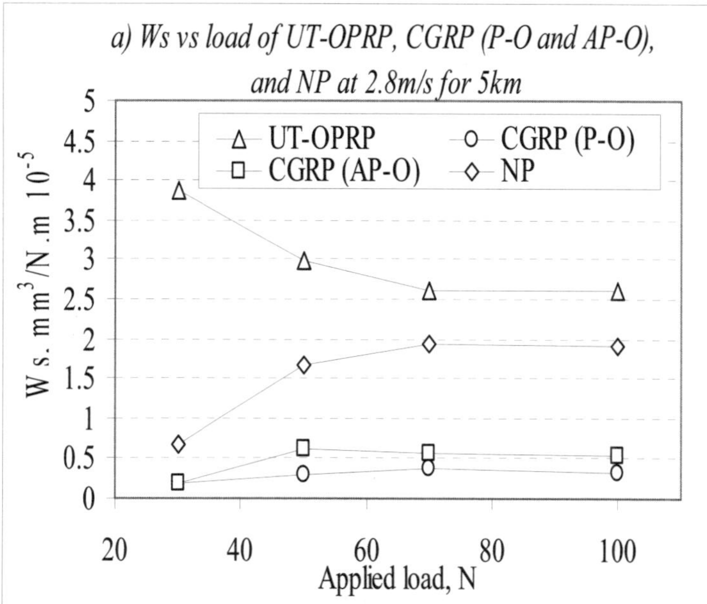

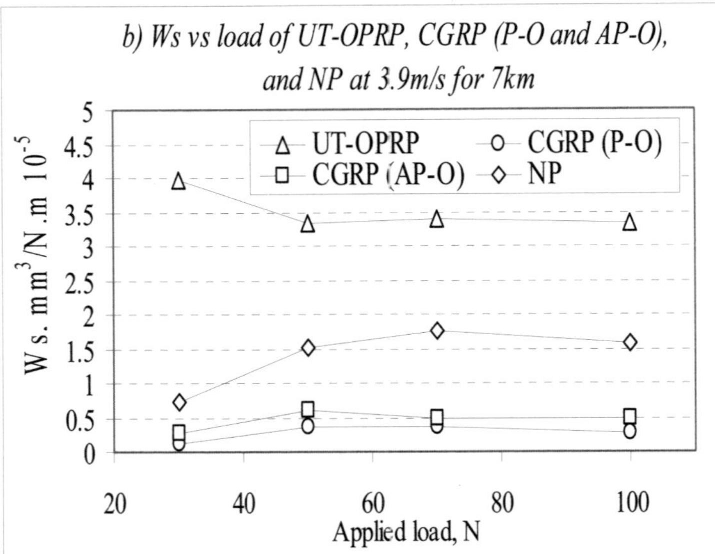

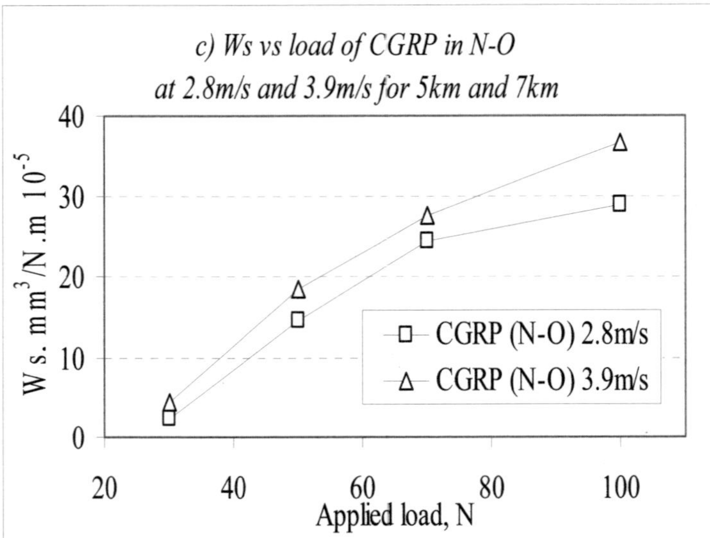

Figure 2.9. *The Ws versus the applied load for neat polyester, UT-PORP and FRPC using the BOR technique under dry contact conditions*

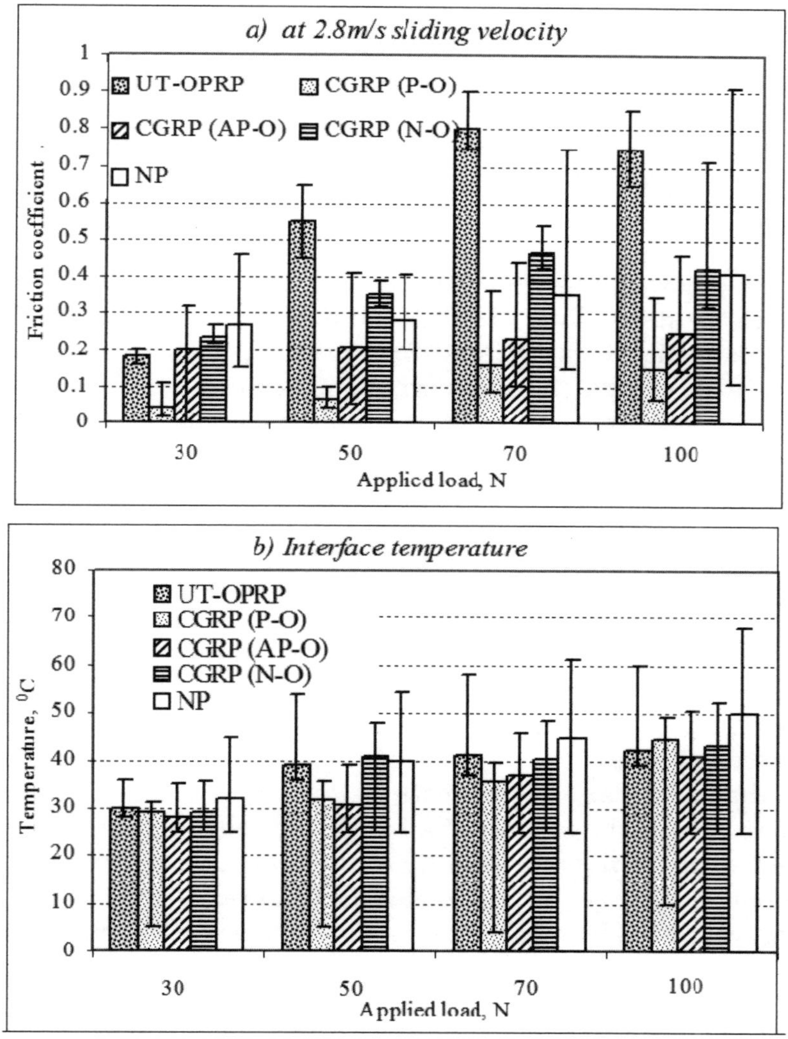

Figure 2.10. *The friction coefficient and the interface temperature versus the applied load of all the materials using the BOR technique under dry contact conditions*

When comparing the tribo-performance of the materials, we can say that UT-PORP shows a lower performance compared to the FRPC (in P-O and AP-O) and neat polyester. In other words, reinforcing the polyester with untreated palm oil fibers worsened the wear performance of the neat polyester. This could be due to the high porosity and poor interfacial adhesion of the untreated palm oil fibers. This may be further clarified by the SEMs of the worn surfaces in Figure 2.11.

At a low applied load of 30 N, Figure 2.11a shows that there is debonding of fibers and the production of fine polyester debris, associated with plastic deformation. This is due to the high thermo-mechanical loading. At an applied load of 50 N, the surface deteriorated, whereas the fine fibers from the bundles were extracted and exposed to the rubbing surface of the composite, see Figure 2.11b. Since there was no sign of back-transfer film on the surface, it is suggested that the extracted fibers and the fine polyester debris were worn away.

At 70 N applied load, it seemed that the separation (gap) between the fibers and the matrix was still not totally filled with polyester debris, which indicates that most of the debris was worn away, see Figure 2.11c. However, the cross-section of the fiber was covered with a layer of polyester. Further increases in the load or the sliding velocity highly softened the polyester regions, leading to them being fragmented into small and large pieces of debris. Figures 2.11d, e and f. Figure 2.11e show that the bundle of fibers was empty and some of the damaged fibers had been worn away.

It may be concluded that with the higher shear at the interface caused significant damage to UT-PORP surface during sliding, which led to a significant removal of material. In other words, the untreated palm oil fibers were weak at the interface and had a poor resistance to shear at high loads and temperatures. This could explain the poor performance of UT-PORP composites.

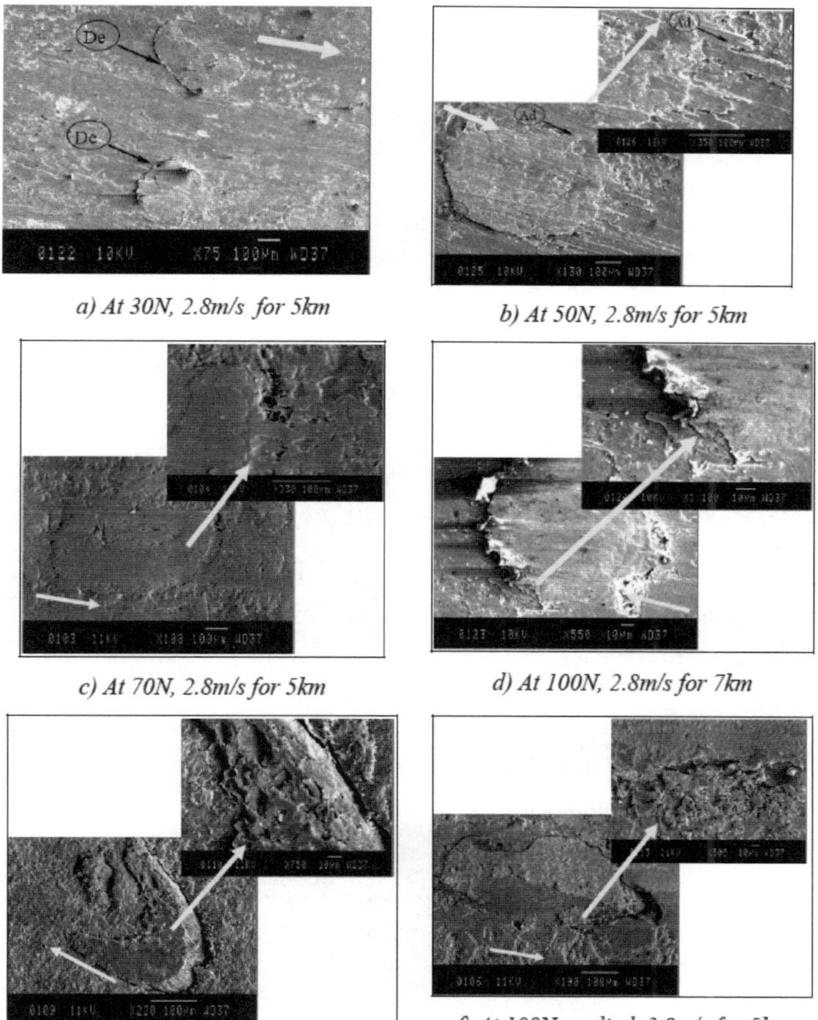

a) At 30N, 2.8m/s for 5km

b) At 50N, 2.8m/s for 5km

c) At 70N, 2.8m/s for 5km

d) At 100N, 2.8m/s for 7km

e) At 70N, 3.9m/s for 7km

f) At 100N applied, 3.9m/s for 5km

Figure 2.11. *Worn surfaces of UT-PORP composite following various tests using the BOR technique under dry contact conditions*

FRPC composite, see Figures 2.12a and 2.12b show that in P-O and AP-O, neither the fibrous (F) nor the resinous (R) regions of the surface were dramatically damaged. There seemed to be a transfer of polyester and fibers from one phase to the other in AP-O. This means that less material was removed, which could explain the high performance of the FRPC composite in the P-O and AP-O. However, the FRPC composite in N-O exhibited poor results, and this was due to the modification that occurred on the composite surface during the rubbing process, i.e. either a layer of polyester or of CSM glass fibers was exposed to the rubbing. Figure 2.12c shows that a layer of polyester has been worn away and that sliding took place on the CSM glass fibers. This may explain why the removal of material in this orientation was high.

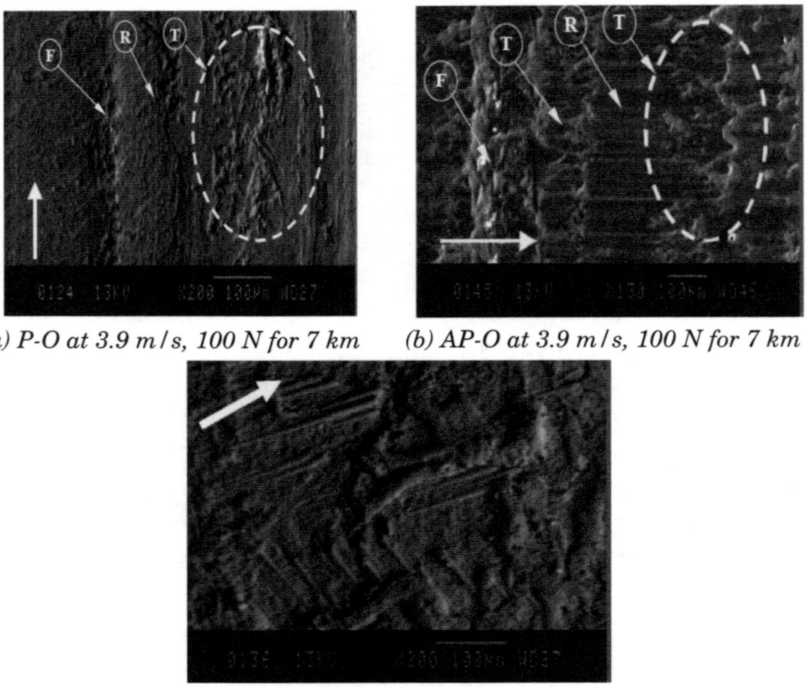

(a) P-O at 3.9 m / s, 100 N for 7 km　*(b) AP-O at 3.9 m / s, 100 N for 7 km*

(c) N-O at 2.8 m / s, 70 N for 5 km

Figure 2.12. *Worn surfaces of the FRPC composite in different orientations using the BOR technique under dry contact conditions [YOU 08]*

2.4.6. *Effect of 6% NaOH treatment*

Figure 2.13 displays the Ws, friction coefficient and interface temperature as functions of the sliding distance for T-PORP, UT-PORP and FRPC (in P-O) at an applied load of 50N with a sliding velocity of 2.8 m/s. This figure indicates that the Ws of the T-PORP composite was lower than that of UT-PORP by about 11%, which is the same as FRPC (in P-O). This is likely to be attributed to an improvement in the interfacial adhesion of the fibers and a reduction in the porosity of the composite.

The worn surface of the T-PORP composite, see Figure 2.14, has no pull-out, bending or breakage of the fibers. At a longer sliding distance, see Figures 2.14b and c, deformation occurred on the surface and the fibers were totally covered with polyester. In the magnified image (Figure 2.14c), we can see that the polyester has been softened, and a transfer film was created.

This transfer film reduced the removal of material. The friction coefficient of the T-PORP composite was lower than that of the UT-PORP composite. This reduces the interfacial heat, i.e. the polyester regions were not significantly softened (see Figure 2.14) as seen before on the UT-PORP composite surface (Figure 2.11).

T-PORP demonstrated a more stable friction coefficient (of about 0.37) compared to the UT-PORP composite (0.42–0.7). The lower friction coefficient of the T-PORP composite could be due to the lower adhesion and interaction between the composite and the counterface asperities.

The above result suggested that the wear and frictional performance of polyester reinforced with treated palm oil fibers was highly competitive with that of glass fibers.

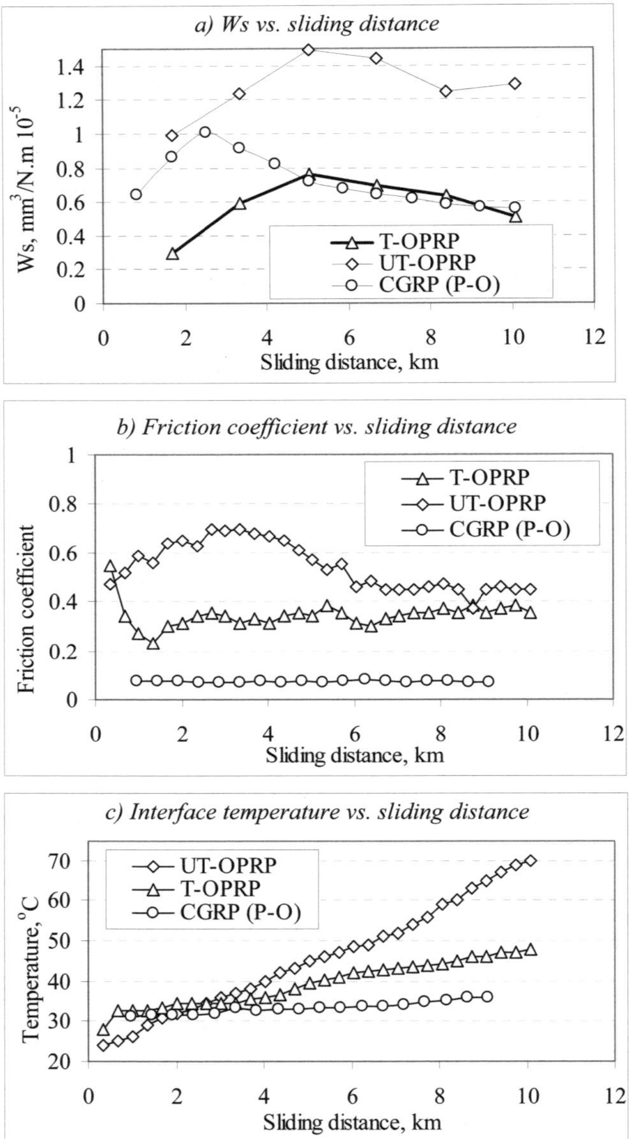

Figure 2.13. *The Ws, friction coefficient and interface temperature versus sliding distance for T-PORP, UT-PORP, and FRPC (in P-O) at 2.8 m/s sliding velocity and an applied load of 50 N using the BOR technique under dry contact condition*

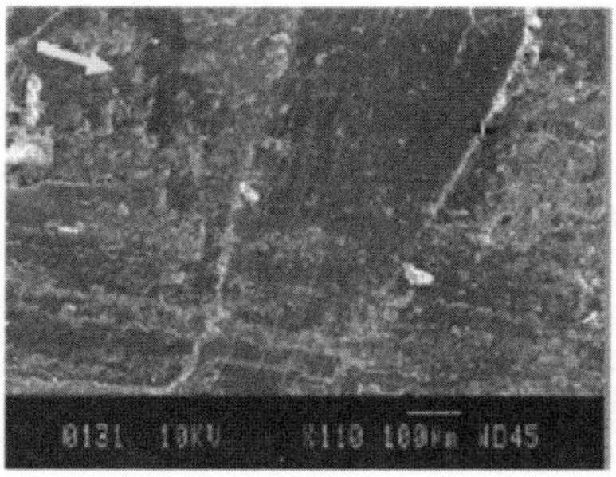

a) For 3.36 km sliding distance

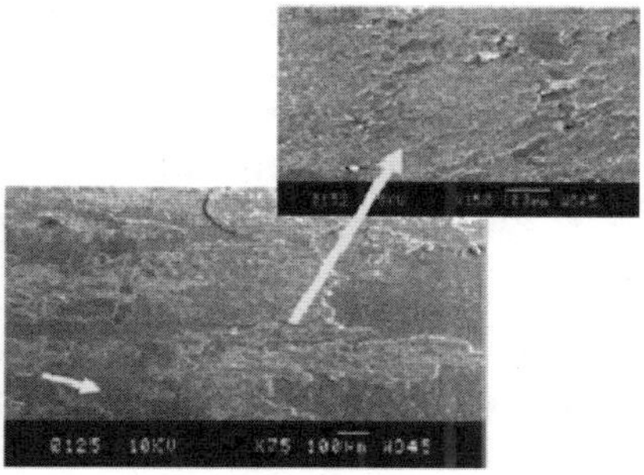

b) For 5 km sliding distance

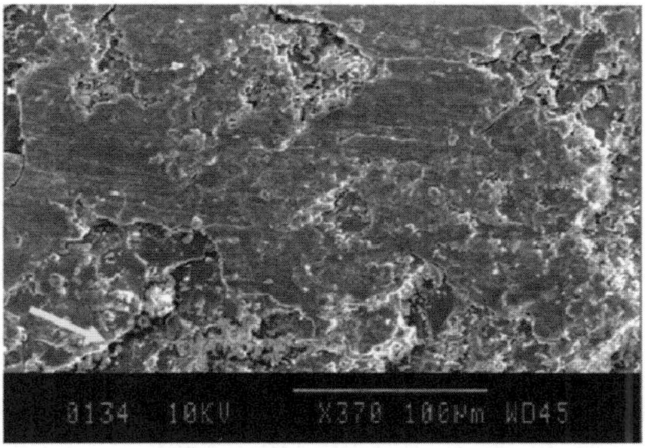

c) For 5 km sliding distance

Figure 2.14. *The worn surfaces of the T-PORP composite tested using the BOR technique under dry contact conditions at 2.8 m/s and 50 N for various sliding distances*

2.4.7. *Effect of wet and dry contact conditions*

Using water as a coolant and cleaner could improve the wear performance of the composite when tested using BOR. The Ws and friction coefficient results of T-PORP, UT-PORP, FRPC (in three orientations) as well as those of neat polyester are presented as function of sliding distance in Figure 2.15. In general, after a 14 km sliding distance, the

Ws remained almost unchanged for all of the materials. The highest wear performance was exhibited by T-PORP. The Ws of neat polyester was reduced by about 60% when reinforced with treated palm oil fibers. The performance of the FRPC in P-O and AP-O was at an intermediate level. The FRPC in N-O and the neat polyester showed very poor results.

The high performance of the T-PORP composite could be due to the reasons given when the same material was tested using BOD, i.e. due to a high interfacial adhesion and low composite porosity. The worn surface of the T-PORP, Figure 2.16a, seemed to be smoothened and the damage appeared due to the removal of the covered polyester layer, without the fibers being torn or damaged. The presence of polyester in the bundle strengthened the fibers and led to a reduction in the amount of material that was removed.

Figure 2.16b shows that the damage that occurred on the UT-PORP composite was more severe for the T-PORP composite. Some of the fibers in the bundles became separated from the surface and were removed. This was due to the poor interfacial adhesion of the fibers and the high porosity of the composite. The relatively poor wear performance of the FRPC under wet contact conditions was due to fragments of the glass fibers acting as a third body (see Figures 2.16c, d and e). This resulted in removal of a considerable amount of material, see Figures 2.16a and 2.16d, as mentioned before in the case of the BOD technique. The worn surface of the FRPC in N-O showed a higher removal of material due to the delamination of numerous layers of CSM glass fibers and the polyester being worn away.

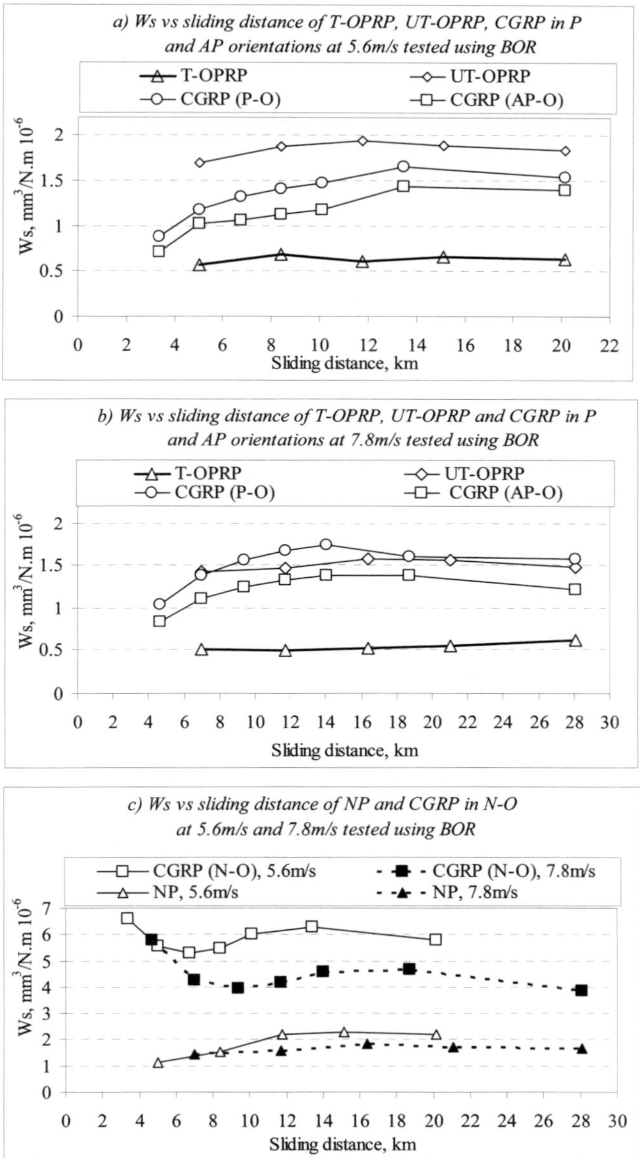

Figure 2.15. *The Ws vs. sliding distance of T-PORP, UT-PORP and FRPC (in P-O, AP-O, N-O) as well as for neat polyester at 50 N applied load tested using the BOR technique under wet contact conditions*

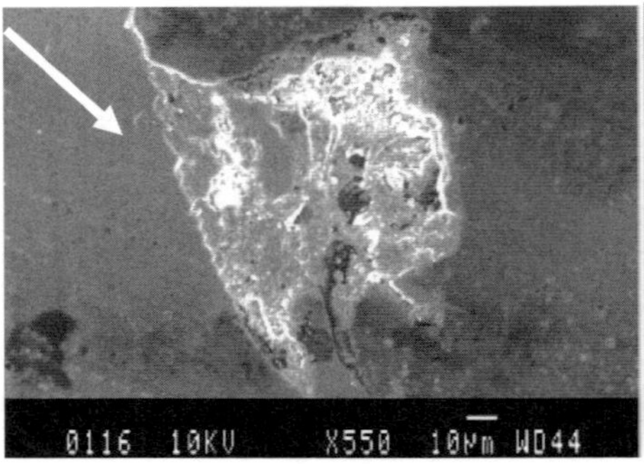

a) T-PORP

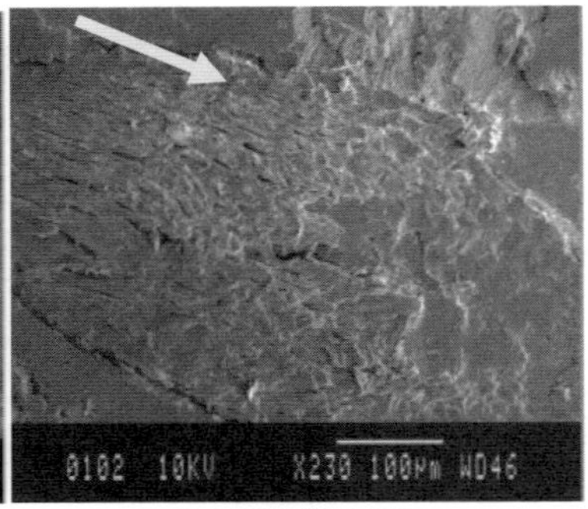

b) UT-PORP

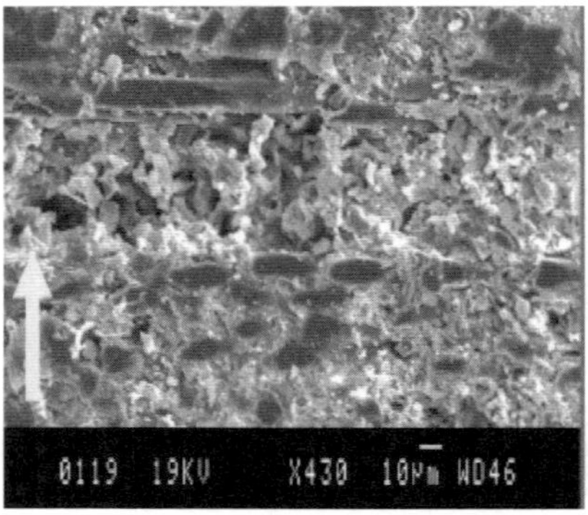

(c) FRPC (AP-O) [YOU 08]

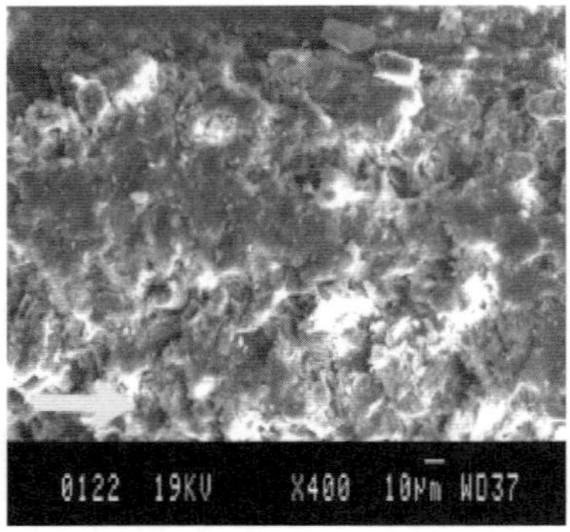

(d) FRPC (P-O) [YOU 08]

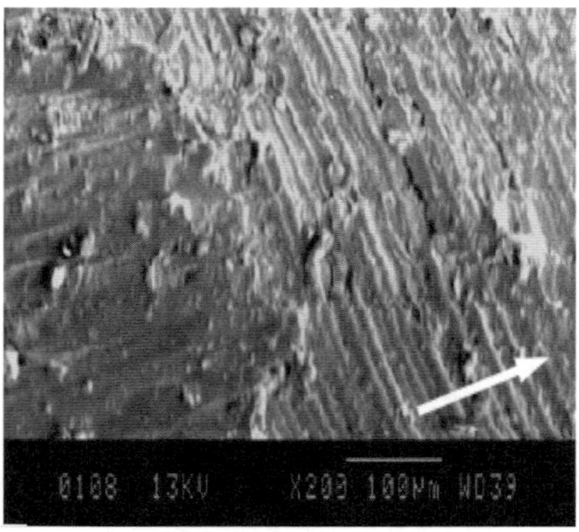

(e) FRPC (N-O) [YOU 08]

Figure 2.16. *SEMs of T-PORP, UT-PORP, FRPC (in P-O, AP-O and N-O) tested using the BOR technique under wet contact conditions*

The average friction coefficient for the materials, see Figure 2.17, shows that the T-PORP composite had the lowest friction coefficient among all of the materials tested. This could be due to be its smooth surface, as previously observed in Figure 2.16a. The highest friction coefficient was found in the FRPC (AP-O) and was thought to result from the rubbing mechanism of the composite in this orientation, where the polyester and CSM glass fiber layers were perpendicular to the sliding direction, see Figure 2.16c.

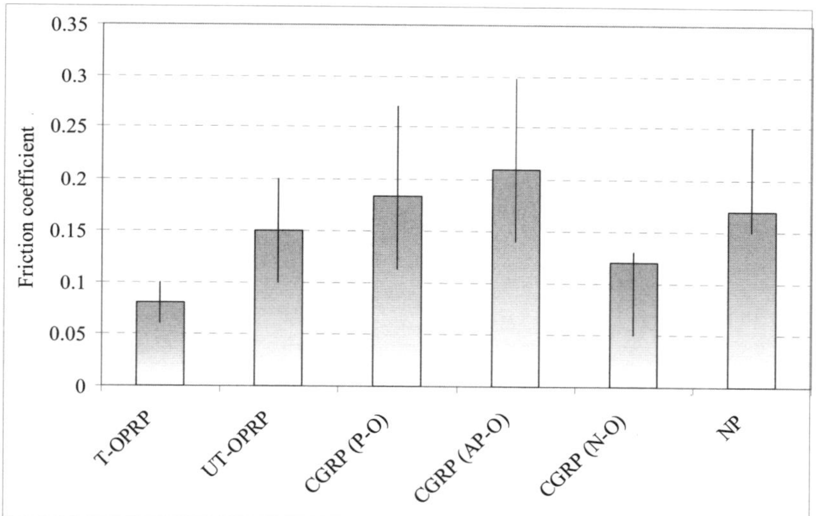

Figure 2.17. *The friction coefficient for the FRPC (in P-O, AP-O, N-O), UT-PORP and T-PORP composites as well as neat polyester at 5.6 m/s sliding velocity and a 50 N applied load when using the BOR technique under wet contact conditions*

2.5. Summary

At variable apparent areas of contact, neat polyester, FRPC (in N-O) gave poor results. Meanwhile, the FRPC in P-O and AP-O offered relatively similar wear performances. Table 2.3 summarizes the adhesive wear trends and Ws ranges of the FRPC (P-O), UT-PORP and T-PORP composites at variable apparent sliding contact areas under wet and dry contact conditions. The T-PORP composite offered the best wear performance under wet and dry contact conditions, compared to the others. FRPC in P-O showed moderate wear results under both wet and dry contact conditions.

Materials	Specific wear rate (Ws), mm^3/N.m × 10^{-5} (BOR)		Wear performance of the composite with respect to others
	Dry	Wet	
FRPC P-O	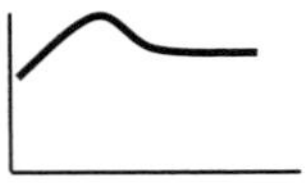 Ws=0.6 to 1	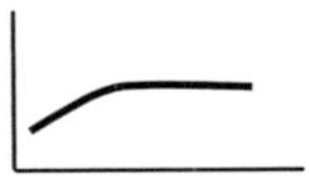 Ws=0.065 to 0.16	Moderate under wet and dry conditions
UT-PORP	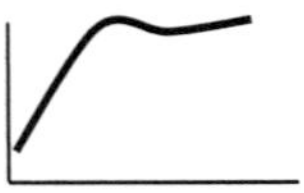 Ws = 1.0 to 1.45	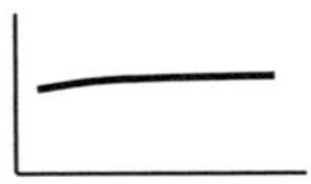 Ws = 0.16 to 0.19	Poor under dry and moderate under wet conditions
T-PORP	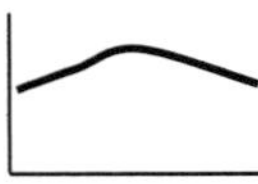 Ws= 0.25 to 0.8	 Ws= 0.05 to 0.06	Best under wet and dry conditions

Table 2.4. *Adhesive wear results of FRPC (in AP-O), UT-PORP and T-PORP composites using BOR under wet and dry contact conditions*

2.6. Bibliography

[ABD 07] ABDELMOULEH M., BOUFI S., BELGACEM M.N., DUFRESNE A., "Short natural-fibre reinforced polyethylene and natural rubber composites: Effect of silane coupling agents and fibres loading", *Composites Science and Technology*, vol. 67, no. 7-8, p. 1627-1639, 2007.

[BRA 05] BRAHMAKUMAR M., PAYITHRAN C., PILLAI R.M., "Coconut fibre reinforced polyethylene composites: effect of natural waxy surface layer of the fibre on fibre/matrix interfacial bonding and strength of composites", *Composites Science and Technology*, vol. 65, no. 3-4, p. 563-569, 2005.

[CHA 07] CHAND N., DWIVEDI U.K., "High stress abrasive wear study on bamboo", *Journal of Materials Processing Technology*, vol. 183, no. 2-3, p. 155-159, 2007.

[CHA 06] CHAND N., DWIVEDI U.K., "Effect of coupling agent on abrasive wear behaviour of chopped jute fibre-reinforced polypropylene composites", *Wear*, vol. 261, no. 10, p. 1057-1063, 2006.

[CHA 07] CHAND N., DWIVEDI U.K., ACHARYA S.K., "Anisotropic abrasive wear behaviour of bamboo (Dentrocalamus strictus)", *Wear*, vol. 262, no. 9-10, p. 1031-1037, 2007.

[DHA 07] DHAKAL H.N., ZHANG Z.Y., RICHARDSO M.O.W., "Effect of water absorption on the mechanical properties of hemp fibre reinforced unsaturated polyester composites", *Composites Science and Technology*, vol. 67, no. 7-8, p. 1674-1683, 2007.

[EIS 95] ELSAYED A.A., ELSHERBINY M.G., ABO-EL-EZZ A.S., AGGAG G.A., "Friction and wear properties of polymeric composite materials for bearing applications", *Wear*, vol. 184, no. 1, p. 45-53, 1995.

[ELT 06] EL-TAYEB N.S.M., YOUSIF B.F., YAP T.C., "Tribological studies of polyester reinforced with CSM 450-R-glass fiber sliding against smooth stainless steel counterface", *Wear*, vol. 261, no. 3-4, 30, p. 443-452, 2006.

[HAR 03] HARSHA A.P., TEWARI U.S., "Two-body and three-body abrasive wear behaviour of polyaryletherketone composites", *Polymer Testing*, vol. 22, no. 4, p. 403-418, 2003.

[HAS 06] HSAHMI S.A.R., DWIVEDI U.K., CHAND N., "Friction and sliding wear of UHMWPE modified cotton fibre reinforced polyester composites", *Tribology letter*, vol. 21, no. 2, p. 79-87, 2006.

[HAS 07] HASHMI S.A.R., DWIVEDI U.K., CHAND N., "Graphite modified cotton fibre reinforced polyester composites under sliding wear conditions", *Wear*, vol. 262, no. 11-12, p. 1426-1432, 2007.

[HAW 99] HAWK J.A.., WILSON R.D., TYLCZAK J.H., DOGAN O.N., "Laboratory abrasive wear tests: Investigation of test methods and alloy correlation", *Wear*, vol. 225-229, no. 2, p. 1031-1042, 1999.

[JAC 04] JACOB M., THOMAS S., VARUGHESE K.T., "Mechanical properties of sisal/palm oil hybrid fiber reinforced natural rubber composites", *Composites Science and Technology*, vol. 64, no. 7-8, p. 955-965, 2004.

[KEN 97] KENNETH G.B., "Resistance to particle abrasion of selected plastic", *Wear*, vol. 203-204, p. 302-309, 1997.

[LUZ 07] LUZ S.M., GONCALVES A.R., DEL'ARCO A.P., "Mechanical behavior and microstructural analysis of sugarcane bagasse fibers reinforced polypropylene composites, Composites Part A:", *Applied Science and Manufacturing*, vol. 38, no. 6, p. 1455-1461, 2007.

[OKU 04] OKUBO K., FUJII T., YAMAMOTTO Y., "Development of bamboo-based polymer composites and their mechanical properties, Composites part A:", *Applied Science and Manufacturing*, vol. 35, no. 3, p. 377-383, 2004.

[TON 98] TONG J., AMELL R.D., REN L., "Dry sliding wear behaviour of bamboo", *Wear*, vol. 221, no. 1, p. 37-46, 1998.

[TON 95] TONG J., REN L., L.I. J., CHEN B., "Abrasive wear behaviour of bamboo", *Tribology International*, vol. 28, no. 5, p. 23-327, 1995.

[TON 05] TONG J.N., MA Y., CHEN D., SUN J., REN L., "Effects of vascular fiber content on abrasive wear of bamboo", *Wear*, vol. 259, no. 1-6, p. 78-83, 2005.

[TOW 08] TOWO A.N., ANSELL M.P., "Fatigue of sisal fibre-reinforced composites: Constant-life diagrams and hysteresis loop capture", *Composites Science and Technology*, vol. 68, no. 3-4, p. 915-924, 2008.

[ZHA 07] ZHANG H., ZHONG Z., FRIEDRICH K., "Effect of fiber length on the wear resistance of short carbon fiber reinforced epoxy composites", *Composites Science and Technology*, vol. 67, no. 2, p. 222-230, 2007.

[YOU 08] YOUSIF B.F., El-TAYEB N.S.M., "Wear and friction characteristics of FRPC composite under wet contact condition using two different test techniques", *Wear*, vol. 265, no. 5-6, p. 856-864, 2008.

Chapter 3

Resistance to Cavitation Erosion: Material Selection

This chapter deals with the use of vibratory cavitation erosion apparatus in evaluating the wear resistance of a material to cavitation erosion in deionized water and selecting materials for applications subject to cavitation erosion. A modified configuration of sample mounting enables us to evaluate the wear resistance of different kinds of materials including metals, alloys, ceramics and polymers under the same test conditions. The comparison of the wear resistances of various bulk materials (metals, alloys, ceramics and polymers) and coatings to cavitation erosion in deionized water is useful for material selection.

3.1. Cavitation erosion of materials – a brief review

The term *cavitation erosion* describes the physical phenomenon of cavitation-induced erosive wear in liquid media [POH 96]. Cavitation is caused by the repeated

---

Chapter written by Jinjun LU, Zhen LI, Xue GONG, Jiesheng HAN and Junhu MENG.

nucleation, growth and violent collapse of clouds of bubbles within a liquid. Damage to material caused by cavitation erosion is widely found in hydraulic machines, such as pumps, hydraulic turbines, valves and ship propellers [HAN 92]. There is no doubt that this kind of material damage must be minimized. Up until now, great efforts have been made to prevent mechanical components experiencing severe material damage subject to cavitation erosion. Despite this, there is still a long way to go.

To understand material damage from cavitation erosion, we should understand micro-jets and their impact on materials. Micro-jets are generated by the implosion of bubbles near a material's surface, resulting in damage created by the high pressures, from about 200 to up to 1,000 MPa and are characterized by high speeds of up to 500 m/s. As a consequence, over time important structural materials are severely damaged due to cavitation erosion. It is therefore of great interest to us to explore materials with high erosion resistance. We want to find the principle for selecting materials with long incubation times and low erosion rates. So far, the cavitation erosion of different kinds of materials has been investigated.

The materials analyzed include:

– metals (e.g. steels [HAT 04], titanium alloys [MOC 07], etc.);

– ceramics (Al_2O_3, SiC, ZrO_2, etc.) [LIT 06, LU 08, TOM 94, TOM 99];

– intermetallic compounds [HOW 95];

– metallic glass [DRO 07];

– polymers [HAT 11].

Surface modification (laser cladding [MAN 03, TAN 06]) is also used to improve the wear resistance of materials. Now it is feasible to measure the wear resistance of a material to cavitation erosion on a vibratory cavitation erosion apparatus, which meets the ASTM Standard G32-06 requirement [AST 06]. In the following sections, material selection based on a vibratory cavitation erosion apparatus is introduced, along with several examples.

3.2. Measuring the wear resistance of a material to cavitation erosion by using a vibratory cavitation erosion apparatus

In a laboratory, a vibratory cavitation erosion apparatus, also called an ultrasonic vibratory apparatus, is the simplest and the most frequently used device for evaluating the wear resistance of a material to cavitation erosion. For an apparatus that meets the requirement of ASTM Standard G32-06, comparison on the data obtained from this apparatus and others is possible and of significance. It is also possible to construct a database on cavitation erosion containing materials from different sources by using a conversion method.

Besides ASTM Standard G32-06, the principle and setup of an ultrasonic vibratory apparatus are well described in many papers. Therefore, the principle and setup of an ultrasonic vibratory apparatus are not discussed in this chapter. Instead, we focus on the basic characteristics of the ultrasonic vibratory apparatus. The methodology and parameters to evaluate the wear resistance of a material to cavitation erosion are also discussed.

3.2.1. *General view of an ultrasonic vibratory apparatus*

In general, an ultrasonic vibratory apparatus has the following characteristics:

– It is simple and inexpensive to design and build.

– The consumption of liquid in a cavitation erosion test is less than that used in other kinds of apparatus.

– The high intensity of the bubble cloud enables it to measure the damage to materials in a relatively short period of time and thereby makes it possible to quickly evaluate and screen materials.

– The temperature and pressure of the liquid can be accurately controlled.

– The size, quantity, and distribution of bubbles cannot be predicted.

– The mounting and shape of a sample have an influence on the experimental results.

– For materials that have a very high or low density, are weak, brittle, or cannot be readily formed into a precisely shaped specimen, there is an alternative method for sample mounting that has been used in recent years, see Figure 3.1.

– Annealed wrought Nickel 200 is suggested as a material for the calibration of an ultrasonic vibratory apparatus. As Nickel 200 is difficult to machine, however, in the newest version of ASTM Standard G32-10 [AST 10], a stainless steel (316) is suggested as an alternative material for calibration.

It is important that a database on the cavitation erosion of materials is constructed, and Hattori and Ishikura started doing just this in 1972. A database on the cavitation erosion of materials can be constructed based on the results obtained on an ultrasonic vibratory apparatus. For the conversion of data, we can refer to [HAT 10]. For example, data other than 50 μm (peak to peak) can be converted into data-values equivalent to 50 μm amplitudes.

The synergetic effect of cavitation erosion and corrosion can be investigated by combining an ultrasonic vibratory apparatus and an electrochemical workstation.

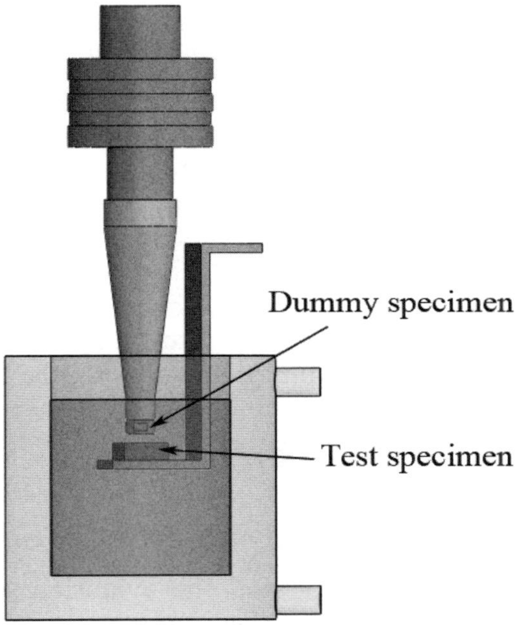

Figure 3.1. *Illustration of a sample mounting for materials that cannot be made into a horn. This type of mounting is used for evaluation in this chapter*

3.2.2. *Determination of the wear resistance of a material to cavitation erosion*

The evaluation of the wear resistance of a material to cavitation erosion using an ultrasonic vibratory apparatus can be found in ASTM standard G32-06. The definitions of terms (incubation time, maximum erosion rate, cumulative mean depth of erosion, etc.) can also be found there. The incubation time, maximum erosion rate and cumulative mean depth of erosion are derived from a cumulative mass loss–time curve. The measurement on the mass of the sample tested should be accurate. In recent years, surface roughness [CHI 05] and area fraction of cavities of the eroded surface [DUL 04] have provided useful information for the incubation stage.

3.2.2.1. *Incubation time, maximum erosion rate and cumulative mean depth of erosion*

The simplest way to evaluate and rank the wear resistance of a material to cavitation erosion is to obtain the incubation time, maximum erosion rate and cumulative mean depth of erosion. A material of high wear resistance to cavitation erosion has a long incubation time or a low maximum erosion rate, or both. The wear resistance of a material to cavitation erosion can be defined as the reciprocal of the cumulative mean depth of erosion. Sometimes the reciprocal of the mean depth of erosion rate is used as the wear resistance of a material to cavitation erosion.

3.2.2.2. *Surface roughness and area fraction of cavities*

Chiu *et al.* have investigated the evolution of the mean surface roughness, R_a, of three common metal materials [CHI 05]. They have found that the roughness–time curve can be divided into three stages.

In addition, it is found that the three stages of change in roughness approximately coincide with the incubation, acceleration and maximum erosion rate stages in mass loss curves. Chiu *et al.* believe that the roughness measurement not only provides information during the incubation stage, but also possesses several advantages over mass loss curves. The evolution of the area fraction of cavities has also been investigated [DUL 04].

3.2.3. *Experimental details*

3.2.3.1. *Ultrasonic vibratory apparatus*

All cavitation erosion tests in this chapter were carried out on a commercially available UVA-1 ultrasonic vibratory apparatus. This apparatus was jointly designed and manufactured by Hangzhou Chenggong Ultrasonic Company and the Lanzhou Institute of Chemical Physics according to ASTM standard G32-06. Table 3.1 summarizes the essential technical data for the experimental setup.

Frequency of the ultrasonic transducer	F = 20 ±0.5 [kHz]
Peak-to-peak amplitude of the test specimen or dummy specimen tip	A = 50 ±2 [μm]
The test (dummy) specimen button diameter	D = 15.9 ±0.02 [mm]
Depth of the specimen immersion	I = 12 ±1 [mm]
Height of the test liquid	H = 100 ±5 [mm]
Distance between the faces of the dummy tip and the test specimen	A = 0.5 ±0.01 [mm]

Table 3.1. *Essential parameters of the UVA-1 for cavitation erosion testing based on the ASTM G32–06 standard*

3.2.3.2. *Cavitation erosion test*

The cavitation erosion tests are conducted in deionized water at a temperature of 25 ± 2°C. The barometric pressure at Lanzhou is 85 ±1 kPa. The illustration of the experimental setup is shown in Figure 3.1. The horn is made of stainless steel and the worn horn should be replaced by a new one every five hours. The sample is 25 mm × 25 mm × 5 mm in size.

The masses of the sample before and after the cavitation erosion test are measured on a balance with an accuracy of 0.1 mg. The mass of the eroded sample is generally measured at 30-minute or 60-minute intervals. For a material with a high wear resistance to cavitation erosion, 120-minute and 240-minute intervals are also used.

After the measurement, a cumulative mass loss–time curve is plotted. In this chapter, the cumulative mean depth of erosion–time curve is used to determine the wear resistance of a material. The cumulative mean depth of erosion (MDE) is calculated by the expression:

$$\mathrm{MDE}(\mu\mathrm{m}) = \frac{10\Delta W}{\rho A}, \qquad [3.1]$$

where:

– ΔW is the mass loss in mg;

– ρ is the density in g cm^{-3};

– A is the cavitated area in cm^2.

To compare the wear resistance of materials that have different densities, the time to a mean depth of erosion of 5 μm, t[5 μm], is used in this chapter.

For a specific material, three parallel tests are conducted. The values in the figures are the average value from the three parallel tests.

3.2.3.3. *Sample preparation*

In this section, the sample preparation for the configuration shown in Figure 3.1 is introduced. For a qualified sample, a parallel surface with low waviness is necessary. This is because otherwise the intensity of the bubbles would vary over the surface due to the distance from the specimen to the horn tip of the transducer not being uniform. Grinding of the sample is helpful to get a parallel surface with low waviness.

Surface finishing of the sample is very important and has a pronounced effect on cavitation erosion resistance. It has been found that the damage and removal of material starts at existing surface defects. The lower the number of surface defects, the higher the wear resistance of a material. The surface defects can be minimized or even eliminated by surface finishing, e.g. polishing.

Material	Chemical composition, wt.%
Pure copper (C11000)	99.99 Cu
Cu-Zn alloy	15.0–17.0 Zn, 0.9–1.1 Si
Cast iron (HT-200)	3.2–3.5 C, 1.5–1.8 Si, 0.8–1.0 Mn, S ≤0.12, P ≤0.15, balance Fe
AISI 1045 steel	0.45–0.5 C, 0.17–0.37 Si, 0.5–0.8 Mn, Cr ≤0.25, Ni ≤0.3, Cu ≤0.25, balance Fe
316L steel	C ≤0.03, Si ≤1.0, Mn ≤2.0, P ≤0.035, S ≤0.03, 12–15 Ni, 16–18 Cr, 2–3 Mo, balance Fe
Hastelloy C-276	14.5–16.5 Cr, 15–17 Mo, 4–7 Fe, 3–4.5 W, 2.5 Co, 0.01 C, 1 Mn, 0.08 Si, 0.35 V, 0.04 P, 0.03 S, balance Ni
ZTA (Al_2O_3-ZrO_2)	10.0 ZrO_2
WC-Co (CG35)	8.0 Co, balance WC
PTFE	100 % polytetrafluoroethylene
UHMWPE	100% ultra-high molecular weight polythylene

Table 3.2. *Chemical composition of the materials*

The wear resistances of 10 commercially available materials are evaluated in this chapter. These materials are classified into three groups: metal and alloy, ceramic and polymer. Their chemical compositions are listed in Table 3.2.

The surface of the sample to be eroded is first ground and then polished on a Phoenix Beta grinder/polisher (made in Buehler, USA). The two polymers, however, are polished using abrasive paper. The surface roughness, R_a, of all of the samples is <0.15 μm.

3.2.3.4. *Corrosion resistance of a material*

The corrosion resistances of the metal and alloys used in this chapter are measured on a CHI660 electrochemical workstation.

3.3. Material selection

There are many materials that can be used in water. Some of them are selected for their availability and low cost. Examples of these are cast iron and AISI 1045 steel. Stainless steel, pure copper and Hastelloy C-276 show very good corrosion resistance in water. Some polymers have received considerable attention and have made a considerable impact regarding in the analysis of cavitation erosion in recent years.

In the following sections, the wear resistances of metal, alloys, ceramics and polymers are investigated and compared. The corrosion resistances of metal and alloys are also measured.

3.3.1. *Metal and alloys*

3.3.1.1. *Cast iron and steel*

The mean depth of erosion – time curves of a cast iron and two steels in deionized water are shown in Figure 3.2. Among these, cast iron shows the poorest wear resistance to cavitation erosion in deionized water. A rapid increase in mean depth of erosion after a 120-minute test indicates the onset of severe wear on the cast iron. The wear resistance of AISI 1045 steel (t[5 μm] = 240 min) is much higher than that of the cast iron (t[5 μm] = 80 min). 316L steel is very resistant to cavitation erosion in deionized water and has a t[5 μm] of 330 minutes.

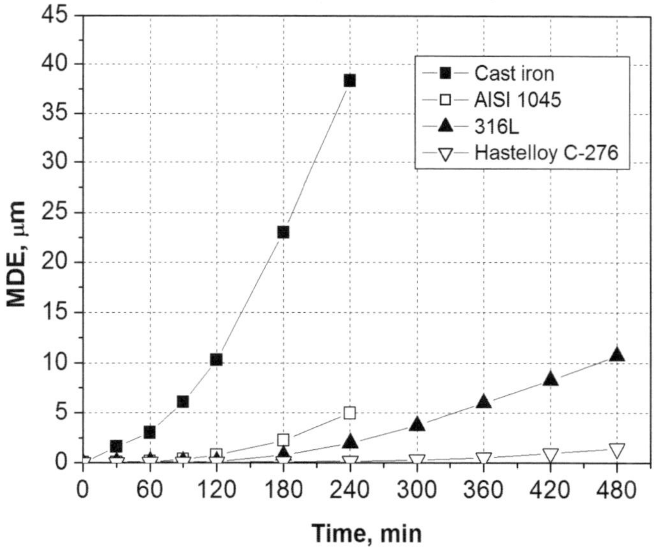

Figure 3.2. *Mean depth of erosion (MDE)–time curves of cast iron, AISI 1045 steel, 316L steel and Hastelloy C-276 in deionized water*

3.3.1.2. *Nickel alloy*

As seen in Figure 3.2, Hastelloy C-276 shows excellent wear resistance to cavitation erosion in deionized water. The t[5 μm] of Hastelloy C-276 is more than 480 minutes and is much higher than that of 316L steel. It indicates that Hastelloy C-276 is a good choice of material when working in water.

3.3.1.3. *Copper alloy*

Pure copper has a very high level of cavitation erosion in deionized water, see Figure 3.3. The t[5 μm] of pure copper is as low as 15 minutes. After a 30-minute test, severe wear of pure copper can be found. The wear resistance of the Cu-Zn alloy is much better than pure copper and even better than cast iron.

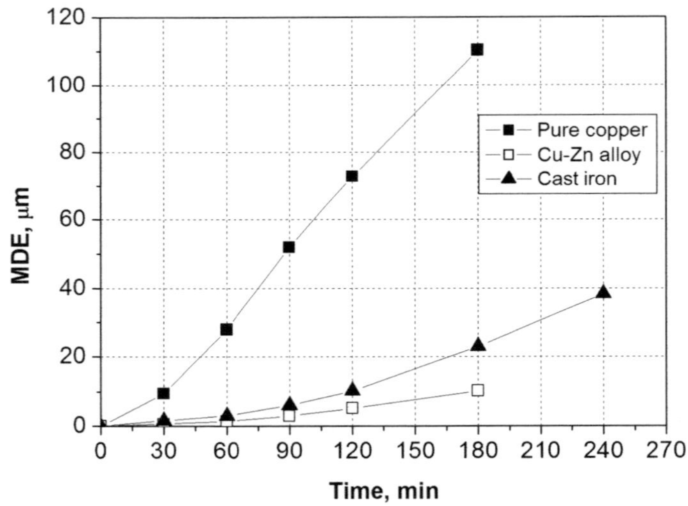

Figure 3.3. *Mean depth of erosion (MDE)–time curves of pure copper and Cu-Zn alloy in deionized water. The cast iron is used as a reference*

3.3.1.4. *Corrosion and wear resistance*

The first point to consider when selecting a material for application in liquid is its corrosion resistance. A material with a high corrosion resistance is preferred. This is true when we combine the results in Figures 3.2 and 3.4. According to Figure 3.4, the ranking of corrosion resistance from low to high in cast iron, AISI 1045, 316L and Hastelloy C-276. The ranking is the same for wear resistance, from low to high, as can be seen in Figure 3.2. However, a controversial result is found for pure copper and Cu-Zn alloy. The corrosion resistance of pure copper is much higher than that of Cu-Zn alloy (see Figure 3.5), while the wear resistance of pure copper to cavitation is much lower than Cu-Zn alloy (see Figure 3.2). It is interesting that pure copper has the highest corrosion resistance but the lowest wear resistance.

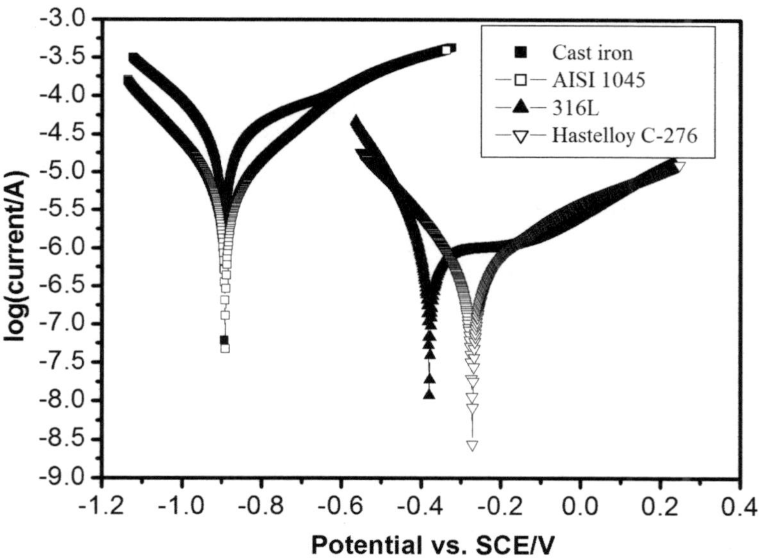

Figure 3.4. *Potentiodynamic polarization curves of cast iron, AISI 1045 steel, 316L steel and Hastelloy C-276 in deionized water (SCE: saturated calomel electrode)*

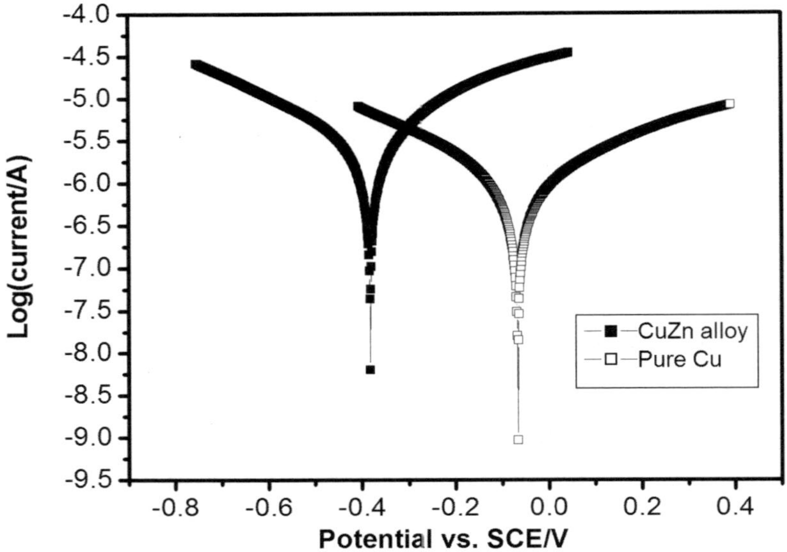

Figure 3.5. *Potentiodynamic polarization curves of pure copper and Cu-Zn alloy in deionized water (SCE: saturated calomel electrode)*

3.3.2. *Advanced ceramic*

Advanced ceramics have good corrosion resistance and this is an advantage over many metallic materials in corrosive liquids. The wear resistance of ZTA ceramic is relatively low when compared to other advanced ceramics, e.g. ZrO_2 and sialon. The t[5 μm] of ZTA is, however, comparable to that of 316L steel (see Figures 3.2 and 3.6). The t[5 μm] of WC-Co is comparable to that of Hastelloy C-276.

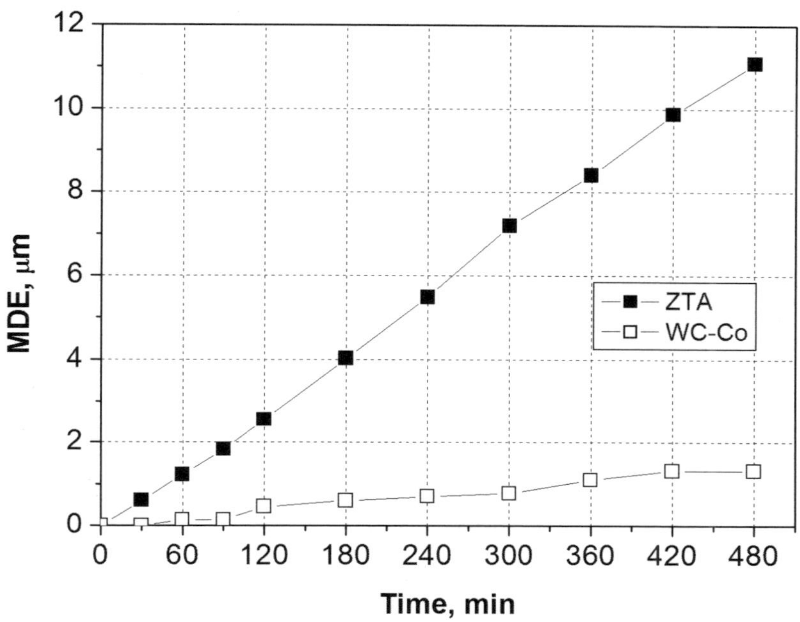

Figure 3.6. *The mean depth of erosion (MDE)–time curves of ZTA ceramic and WC-Co in deionized water*

3.3.3. *Polymer*

PTFE is a soft polymer that can be used as a solid lubricant. The wear resistance of PTFE is comparable to that of cast iron in deionized water. UHMWPE shows a surprisingly long incubation time of >20 hours. In addition, the t[5 μm] of UHMWPE is as high as 1,320 minutes. All of the data indicate that the wear resistance of UHMWPE is the highest from among the materials examined in this chapter. Large strains before fracture might be the main reason for the high wear resistance of UHMWPE to cavitation erosion in deionized water.

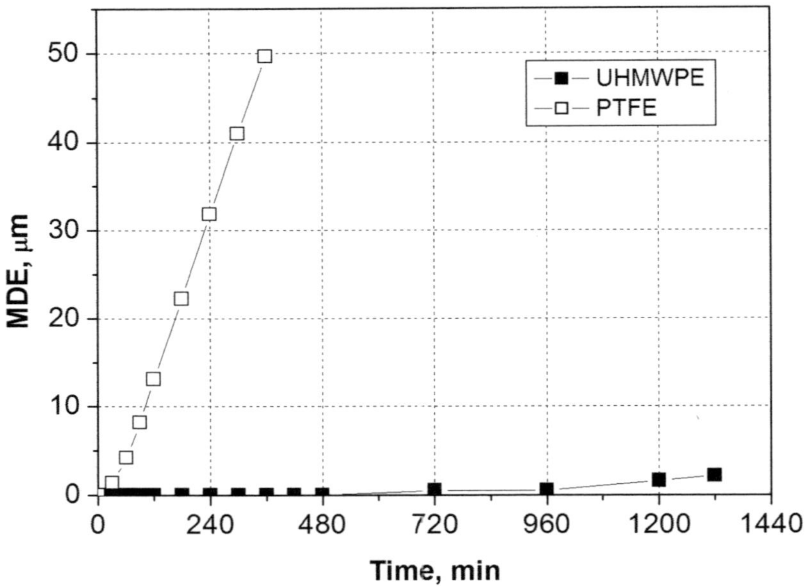

Figure 3.7. *mean depth of erosion–time curves of PTFE and UHMWPE in deionized water*

3.3.4. *Comparison*

In section 3.3, the wear resistances of different materials to cavitation erosion in deionized water are evaluated using an ultrasonic vibratory apparatus. The mean depth of erosion–time curves of all of the materials tested are merged into one graph (Figure 3.8). It can be seen that the wear resistance varies greatly from one material to another. UHMWPE, Hastelloy C-276, WC-Co and 316L are materials with good wear resistance to cavitation erosion in deionized water based on the wear resistance in Figure 3.8.

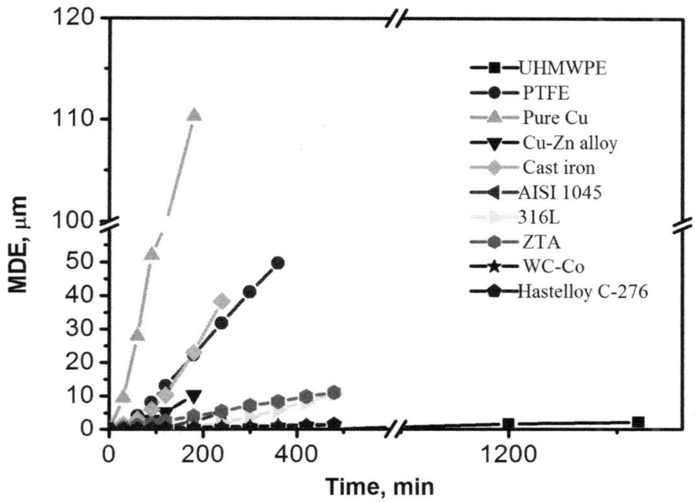

Figure 3.8. *The mean depth of erosion (MDE)–time curves of all of the materials tested in deionized water*

Material selection is not as simple as this, however, in real applications. Besides this, the impact of wear resistance, mechanical strength, corrosion resistance, availability and machinability of a material on the resistance of cavitation erosion should be considered. For example, the machinability of Hastelloy C-276 is not good, and this is the main obstacle to its use in various applications. Although UHMWPE has a high wear resistance and corrosion resistance, the material is very expensive and so is not cost effective.

3.4. Conclusion

An ultrasonic vibratory apparatus can be used to evaluate and rank the wear resistance (t[5 µm]) of different kinds of materials, including metallic materials, ceramics and polymers under the same test conditions.

Steel 316L and Hastelloy C-276 show good wear and corrosion resistance to cavitation erosion in deionized water. Pure copper has excellent corrosion resistance but very poor wear resistance in the same conditions. The wear resistance of WC-Co is comparable to that of Hastelloy C-276 in deionized water.

The material with the highest wear resistance to cavitation erosion in deionized water is UHMWPE. Its incubation time is much longer than that of the rest of the materials tested.

Besides the wear resistance, the mechanical strength, corrosion resistance, availability and machinability of a material to resist cavitation erosion should be considered when selecting a material.

3.5. Acknowledgement

The authors acknowledge and thank Mr Yuanping Chen of Hangzhou Chenggong Ultrasonic Company for his contribution in building and manufacturing the ultrasonic vibratory apparatus.

3.6. Bibliography

[AST 06] ASTM G32-06, "Standard method of vibratory cavitation erosion test", *Annual Book of ASTM Standards*, section 3, vol. 2, ASTM, Philadelphia, USA, 2006.

[AST 10] ASTM G32-10, "Standard method of vibratory cavitation erosion test", *Annual Book of ASTM Standards*, section 3, vol. 2, ASTM, Philadelphia, USA, 2010.

[CHI 05] CHIU K.Y., CHENG F.T., MAN H.C., "Evolution of surface roughness of some metallic materials in cavitation erosion", *Ultrasonics*, vol. 43, no. 9, p. 713-716, 2005.

[DRO 07] DROZDZ D., WUNDERLICH R.K., FECHT H-J., "Cavitation erosion behaviour of Zr-based bulk metallic glasses", *Wear*, vol. 262, no. 1-2, p. 176-183, 2007.

[DUL 04] DULIAS U., ZUM GAHR K-H., "Cavitation erosion of advanced ceramics as function of medium and surface quality", *Materials Science and Engineering Technology*, vol. 35, p. 801-808, 2004.

[HAN 92] HANSSON C.M., HANSSON I.L.H., *Cavitation Erosion ASM Handbook*, vol. 18, ASM International, Materials Park, USA, 1992.

[HAT 04] HATTORI S., ISHIKURA R., ZHANG Q., "Construction of database on cavitation erosion and analyses of carbon steel data", *Wear*, vol. 257, no. 1-2, p. 1022-1029, 2004.

[HAT 10] HATTORI S., ISHIKURA R., "Revision of cavitation erosion database and analysis of stainless steel data", *Wear*, vol. 268, no. 1-2, p. 109-116, 2010.

[HAT 11] HATTORI S., ITOH T., "Cavitation erosion resistance of plastics", *Wear*, vol. 271, no. 7-8, p. 1103-1108, 2011.

[HOW 95] HOWARD R.L., BALL A., "The solid particle and cavitation erosion of titanium aluminide intermetallic alloys", *Wear*, vol. 186-187, no. 1, p. 123-128, 1995.

[LIZ 06] LITZOW U., ZUM GAHR K.-H., Schneider J., "Cavitation erosion of advanced ceramics in water", *International Journal of Materials Research*, vol. 97, p. 1372-1377, 2006.

[LU 08] LU J-J., ZUM GAHR K-H., SCHNEIDER J., "Microstructural effects on the resistance to cavitation erosion of ZrO_2 ceramics in water", *Wear*, vol. 265, no. 11-12, p. 1680-1686, 2008.

[MAN 03] MAN H.C., CUI Z.D., YUE T.M., CHENG F.T., "Cavitation erosion behavior of laser gas nitrided Ti and Ti6Al4V alloy", *Materials Science and Engineering A*, vol. 355, no. 1-2, p. 167-173, 2003.

[MOC 07] MOCHIZUKI H., YOKOTA M., HATTORI S., "Effects of materials and solution temperatures on cavitation erosion of pure titanium and titanium alloy in seawater", *Wear*, vol. 262, p. 522–528, 2007.

[POH 96] POHL M., "Cavitation erosion", *Practical Metallography*, vol. 33, p. 168-187, 1996.

[TAN 06] TANG C.H., CHENG F.T., MAN H.C., "Laser surface alloying of a marine propeller bronze using aluminium powder: Part I: Microstructural analysis and cavitation erosion study", *Surface and Coatings Technology*, vol. 200, no. 8, p. 2602-2609, 2006.

[TOM 94] TOMLINSON W.J., MATTHEWS S.J., "Cavitation erosion of structural ceramics", *Ceramic International*, vol. 20, no. 3, p. 201-209, 1994.

[TOM 99] TOMLINSON W.J., KALITSOUNAKIS N., VEKINIS G., "Cavitation erosion of aluminas", *Ceramic International*, vol. 25, no. 4, p. 331-338, 1999.

Chapter 4

Cavitation of Biofuel Applied in the Injection Nozzles of Diesel Engines

It is necessary to investigate increasing the antiwear and friction reduction abilities of diesel engine nozzles and improving the effects of mixing and atomization of fuels based on the demands of alternative engine fuels. We should study the origin of cavitation erosion on the nozzles of diesel engines when using a biomass-fuel and its relation with cavitation flow, composition and its physical and chemical properties. We also need to optimize the working conditions of biomass fuels to improve diesel and tribological performance. This chapter deals with the introduction of biomass fuels and the effects of cavitation. It is mainly focused on describing the cavitation flow of biofuels and its effects on cavitation erosion in the injections nozzle of diesel engines. The cavitation wear mechanism is also discussed in the case of biofuels as an alternative engine fuel. This chapter helps us to understand how to improve the antifriction and

---

Chapter written by Hengzhou WO, Xianguo HU, Hu WANG and Yufu XU.

antiwear performances of diesel engine nozzles. It also provides a theoretical foundation and guiding principle for the tribological design and selection of nozzle materials when using a biomass-fuel in a diesel engine.

4.1. Introduction

Cavitation happens when liquid (water or fuel) pressure is lowered below its vapor pressure or the vapor pressure is increased due to liquid pressure. This often happens at extremely high speeds. Cavitation occurs inside a pump, a nozzle or around an obstacle, such as a rapidly spinning propeller or in a body of liquid (such as a kettle) due to temperature and/or pressure change. The pressure of the fluid can drop:

– due to its high speed of movement (Bernoulli's principle);

– when the pressure drops below the vapor pressure of the liquid;

– when the temperature increases.

Thus, when vapor pressure increases to reach liquid pressure, it vaporizes – typically forming small bubbles of liquid vapor with its gas phase.

Cavitation is, in many cases, an undesirable occurrence. In devices such as propellers, nozzles and pumps, cavitation causes a great deal of noise, damage to components, vibrations and a loss of efficiency. When the cavitation bubbles collapse, they force energetic liquid into very small volumes, thereby creating spots of high temperature and emitting shock waves, the latter of which are a source of noise.

The noise created by cavitation erosion is a particular problem for military submarines, as it increases the chances of them being detected by passive sonar. As the collapse of a cavity is a relatively low-energy event, however, highly localized collapses can erode metals such as steel over time. The pitting caused by the collapse of cavities produces significant wear on components and can dramatically shorten a propeller, nozzle's or pump's lifetime.

After a surface is initially affected by cavitation, it tends to erode at an accelerating pace. The cavitation pits increase the turbulence of the fluid flow and create crevasses that act as nucleation sites for additional cavitation bubbles. The pits also increase the component's surface area and leave behind residual stresses. This makes the surface more prone to stress corrosion. The wear damage created by cavitation erosion is also a particular problem for the atomization effect of fuels in engine nozzles. This is because cavitation through the nozzle of a diesel injector is known to have a dramatic influence on the liquid jet atomization.

Due to increasing oil prices and concerns over climate change, biomass fuels have become more important as alternative energy sources. It is an open question as to which types of biofuels have the best yield potentials, characteristic properties and environmental consequences for providing the greatest contribution to future energy requirements. Apart from the quality aspects, the question of quantity is very important, i.e. the yields of raw materials from agriculture and forestry for biomass as well as the conversion efficiencies/yields of the conversion process to automotive fuels [SCH 08]. The biofuel will be derived from a process that can convert a variety of waste materials into aviation fuel and can provide life-cycle greenhouse gas savings of up to 95% compared to the fossil fuel-derived jet kerosene.

The transformation of biomass into fuel and chemicals is becoming increasingly popular as a way to mitigate global warming and diversify energy sources. Biomass is a renewable, carbon-neutral resource, and fuels derived from biomass usually burn more cleanly than fossil fuels. It has been estimated that biomass could provide about 25% of the world's energy requirements.

In addition, biomass can also be a source of valuable chemicals, pharmaceuticals and food additives. Several kinds of biomass can be converted to fuel and chemicals. Examples include:

- wood and wood waste;
- agricultural crops;
- agricultural waste;
- litter from animal feedlots;
- waste from food processing operations;
- sludge from water treatment plants.

There are many articles about cavitation with water, but it is necessary to study the effects of variations in fuels on cavitation in the engine nozzle.

4.2. General understanding of cavitation erosion

4.2.1. *Mechanism of cavitation erosion*

Cavitation occurs when pressure drops below the saturated vapor pressure of the liquid, consequently resulting in the formation of gas- or vapor-filled bubbles [BRE 05, HAM 80, OKA 94]. The mechanism of cavitation has been studied for decades through the use of bubble dynamics.

In 1917, Lord Raleigh showed that with the decreasing radius of a bubble, the velocity and acceleration of the bubble's wall increased sharply when the bubbles collapsed. He obtained a theoretical expression for pressure development at a relatively moderate distance outside the spherical boundary of a collapsing bubble [RAY 17]. According to this theory, the fluid pressure could become extremely high when bubble collapse was near completion, a process known as “implosion”.

Figure 4.1 shows some high-speed images that present the dynamics of cavitation bubbles within 600 μs. The position of the focal region is illustrated in Figure 4.1a, with the pointer indicating the direction of ultrasound transmission [CHE 06].

The present understanding of the mechanism of cavitation erosion is far from complete, because it is a complex and highly localized phenomenon that involves the interaction of hydro-dynamical, mechanical, metallurgical and chemical effects. To understand such complex phenomenon, we need a full clarification of the erosion mechanism and controlling parameters inherent to it, so that cavitation erosion can be prevented, or at least reduced. This cannot be done without full information and the collection of massive amounts of data under different conditions.

The most important problem to be solved before a better understanding of cavitation erosion can be achieved is the relationship between erosion damage and the magnitude and frequency of cavitation loading on the surface [MOM 95, WAT 96].

It has long been established that the pits observed in soft material during the early stages of the incubation period are produced by individual cavitation bubble collapses [CHE 08a, WAT 96]. In this initial stage the pits, which are deformed plastically, are produced. From this it is possible to estimate cavitation from the volume displaced and the energy used.

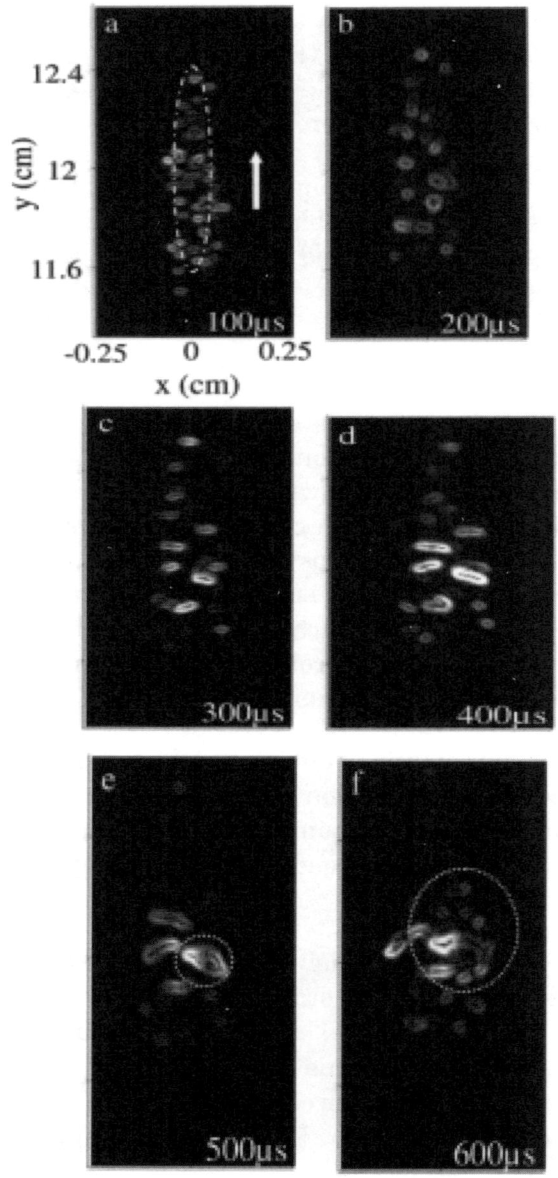

Figure 4.1. *High-speed images showing the dynamics of cavitation bubbles*

4.2.1.1. *Shock-wave theory*

As discussed previously, a collapsing cavity is thought to be responsible for the damage caused by cavitation. During the early years of study pertaining to cavitation damage, based upon Raleigh's computation, it was assumed that cavitation bubbles – as long as they remain spherical – collapse with sufficient force to produce pressures (shock waves) high enough to cause damage, as shown in Figure 4.2 [CHE 08b].

However, it was later observed that cavities in flow conditions that are typically associated with real-life scenarios do not collapse spherically. This implies that the pressures associated with these unsymmetrical collapsing cavities are too low to cause damage.

Even in the case of spherical collapse, it has been shown that the bubble must be so close to the affected surface that the probability of a large number of damaging events is small.

Hickling and Plesset showed that in order to cause damage bubbles must be well within a distance of one bubble's radius from the affected strata [PLE 71]. Even a cavity that is initially spherical becomes unstable and losses its symmetry during collapse.

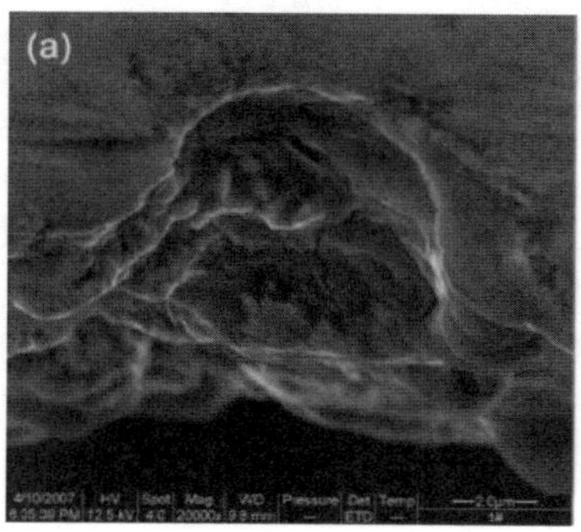

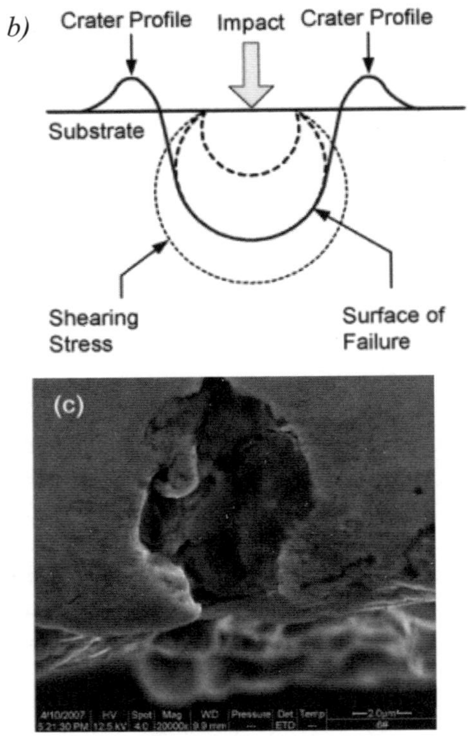

Figure 4.2. *(a) a cross-section of craters; (b) the distribution of shear stress and crater formation; and (c) a cross-section of a crater*

4.2.1.2. *Micro-jet formation theory*

A similar phenomenon can be observed with bubbles in the path of an ultrasonic pressure wave. The impressed acoustic pressure field causes forced oscillations of the bubbles, leading to instability and disintegration. It was initially postulated that the disintegration of existing bubbles led to the formation of jets caused by the flow of water between the new bubbles. It was these jets that were thought to cause the damage.

Given the randomness of the process, however, it was difficult to firmly state that all jets formed in this way would be directed towards the surface under consideration in order to cause damage. The micro-jet formation theory was, however, later proved to be true by Benjamin and Bllis [BEN 66].

The jet formation is depicted in Figure 4.3 [STE 02].

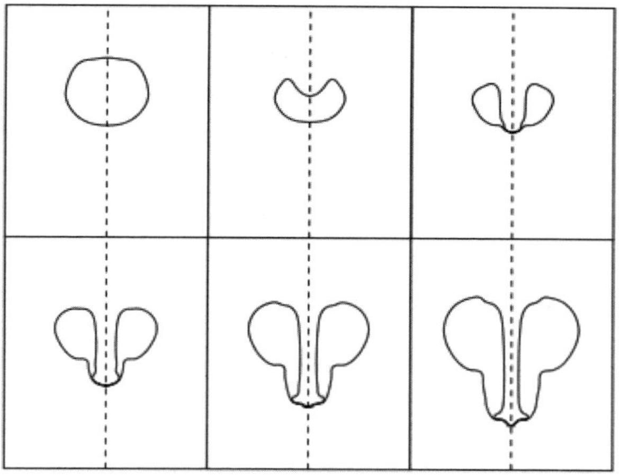

Figure 4.3. *Jet formation and impact*

The jet is formed inside the bubble when its radius is minimal. It can be so strong, however, that it persists, even through the growth cycle of the bubble. Although the discovery of this fact cleared up the doubt surrounding the formation mechanism of jets, engineers were still unsure whether the jet formed due to a collapsing bubble was capable of inflicting damage on a nearby boundary in a flowing system.

The issue raised skepticism about the mechanism by which bubbles are driven close enough to the wall for the jet to cause damage. This issue could be addressed by the fact that the bubbles were attracted to the wall due to the pressure reduction between itself and the wall and the Bjerknes' force arising from the collapsing motion of the cavity.

Another factor that may drive the cavities towards the wall is the application of an external force or pressure gradient. However, the effects of the external stimuli must be in the right balance. If the magnitude of the external forces is not right, the cavity can collapse without having enough time to move close enough to the boundary for the jet to have an effect. Moreover, the orientation of the jet should be normal to the boundary for maximum effect. This could be facilitated by the direction of flow or existing pressure and velocity gradients relative to the rigid boundary in the collapsing zone.

4.2.1.3. *Thermal effect*

It has been acknowledged that the mechanical damage process of cavitation erosion is often accompanied by a thermal effect [HAM 80]. Numerical results acquired by Plesset [PLE 77] and experimental phenomenon observed by Nowotny [NOW 47] and Gavranek *et al.* [GAV 60] have shown that a high temperature was reached in the cavity at the end stage of bubble collapse.

The temperature at bubble collapse is so high that the metal strength is reduced and sometimes the surface melts. Numerical results recently given by Wu and Robert [WU 93] and Ying and An [YIN 02] show that the temperature in the cavity reaches 10,000°C at the moment of collapsing. The effect of temperature changes the physical properties of liquid and as a result the formation and collapse of cavities is enhanced or inhibited.

Zhou and Hammitt [ZHO 83] investigated the effect of temperature on the erosion characteristics of S31600 stainless steel under cavitating condition and found that there is a temperature at which the erosion rate reaches a maximum – 60°C.

4.2.2. *Synergistic effect of cavitation erosion and corrosion*

In comparison with traditional fuels, such as diesel, the use of biofuels in internal combustion engines offers several advantages in terms of logistics as well as the power-to-heat ratio and efficiency, however there is also a new challenge involved, due to the corrosion effect of biofuels. The cavitation erosion that occurs is concentrated at the nozzle orifice, and the mechanism of cavitation erosion is obvious. When biofuel is used as an engine fuel, after a couple days in service the results are beyond expectation [ASI 06] – it has been observed that the wear due to biofuel was greater. If the corrosion or the cavitation erosion was the only dominant mechanism, this should not happen. It is therefore possible that the damage to the nozzles is caused by a complex mechanism of interaction that combines corrosion and cavitation erosion, which makes the process leading to failure easier and faster. In this case, our understanding of the mechanism of cavitation erosion and corrosion is very important if we want to lay a solid foundation for the use of biofuel in diesel engines.

The cavitation damage is actually caused by the combination of mechanical and corrosion effects in the corrosive medium. Related researches indicate that the combination of the two effects has a greater impact than each single action on damage to the material.

4.2.2.1. *Promoting effect of cavitation erosion to corrosion*

First, in the process of cavitation, deformation of the working surface occurs, which induces the appearance of micro-cracks and micro-pores. This process dramatically increases the reactivity, accelerates the process of corrosion due to reactants, and the production of reaction products in the mass transfer process.

Second, the substances (such as oxygen) in the gas phase accelerate the corrosion process by increasing the surface reaction activity and reducing the charge transfer resistance.

Third, the cavitation process creates spots of high local temperature and pressure that make the corrosion reactions easier and faster. During the collapse of bubbles, a large pulse of stress is generated and this is often accompanied by a high-velocity jet of liquid that impinges the solid surface in contact with the liquid. This destroys the protective oxide film, which will obviously accelerate the corrosion.

As the collapse continues, when mechanical or mechanics effects impact the material surface, a high stress concentration will be generated in a tiny region between sub-surface and surface. This will cause a local elastic–plastic deformation of the surface. The internal energy will change sharply in this region, and reach a high level. Therefore, the material in deformed region and the un-deformed region will experience a large difference in internal energy. According to the electrochemical corrosion theory, this and the surrounding area will engender intensive electrochemical non-uniformity, i.e. an electrochemical electric dipole will be formed due to the difference between the electrochemical properties of the material surface.

In the condition of high and repeated impact that is induced by bubble collapse, the function of this electrochemical electric dipole will remain and keep developing. Apparently, this constitutes a vicious circle and results in severe material damage in this region. There is a serious loss of material and the formation of numerous cavitation pits.

4.2.2.2. *The promoting effect of corrosion to cavitation erosion*

In the case of liquid biofuel, corrosion behavior plays a central role in the failure of materials [CHI 03]. Corrosion roughens the material surface, especially in the area of the defects. The emergence of localized corrosion results in many stress concentration points. This accelerates the formation and development of fatigue cracks, and thus promotes the process of cavitation.

As is well known, the work-hardening of a material is close to that of cavitation erosion. The soft metals are prone to cavitation erosion. The process of corrosion can dissolve the work-hardening layer covering the surface of the material and decrease its fatigue strength. This produces the perfect conditions for the emergence and development of cavitation erosion.

4.3. Hydraulic characteristics of cavitation flow

Nozzles play a very important role in the engine fuel injection system. Former studies have indicated:

– the geometric shape of nozzles (such as their inlet radius, length-to-diameter [L/D] ratio and so on);

– the needle lift;

– the working condition (such as injection pressure);

– the properties of fuel;

have obvious influences on fuel flow in nozzles. Cavitation bubbles produced when the local pressure in the flowing nozzle was lower than the vapor pressure of fuel inside the nozzles.

Cavitation mainly results in two kinds of influences:

– one promotes the fuel atomization, which causes complete combustion and reduces emissions;

– the other causes material erosion, thus decreasing the useful life and performance of nozzles.

It is of vital significance that we study the cavitation phenomenon and erosion mechanism in the case of the practical applications of alternate fuels.

The formation of cavitation in diesel injector nozzles is primarily ascribed to the breakdown of the non-equilibrium of liquid dynamics and the separation of boundary layers due to the geometric characteristics of nozzles [LIU 08]. In the modern direct injection diesel engine, the injection pressure is higher. Cavitation is liable to occur when the fuels pass through the passage of nozzles.

Several studies have shown that the cavitation flow in injector nozzles has a strong impact on the spray and atomization characteristics of nozzles and lifespan [PAY 04, SCH 97a]. Cavitation and cavitation erosion in diesel injector nozzles has been the focus of numerous investigations in recent years.

Experimental investigation and numerical simulation are the usual method by which to study cavitating flow in nozzles. Since the size of nozzle used has a diameter of a fraction of millimeters, large-scale nozzles are often used and observed in experiments [GAN 01, PAR 08, SAT 01, SUH 08a]. However, Chaves revealed that there is difference between large-scale and real-size nozzles due to scale effect [CHA 95]. With the development of technology, real-size nozzles have been observed [BAD 99, LIV 08, TAK 05] but this is difficult to carry out. In recent years, with the development of computer technology, computational fluid dynamics has become a universal method by which to simulate cavitation in real-size nozzles [GIA 04, GIA 07, SCH 99].

4.3.1. *Numerical models and validation*

Two-dimensional transient simulation using a homogenous mixture flow model and full cavitation model were carried out [SIN 02]. Isothermal nozzle flow is assumed and there are no velocity slips between the liquid and vapor phase and no consideration is given to energy exchange between the two phases. Simulation results are validated and compared with Hu's experimental results under the same conditions [HU 11]. We can draw the conclusion that the numerical model in the present study is feasible and accurate enough to forecast cavitation in nozzles.

4.3.2. *Effect of boundary pressure on cavitation*

The geometry model used is of a simplified axial sac-type volume nozzle with a single circular orifice. The nozzle has a rounded inlet with a diameter of 0.45 mm, the radius of sac-type volume is 1 mm and the rounded radius is 30 μm. The nozzle length over nozzle diameter (L/D) ratio is 4. Figure 4.4 shows the boundary conditions for nozzles.

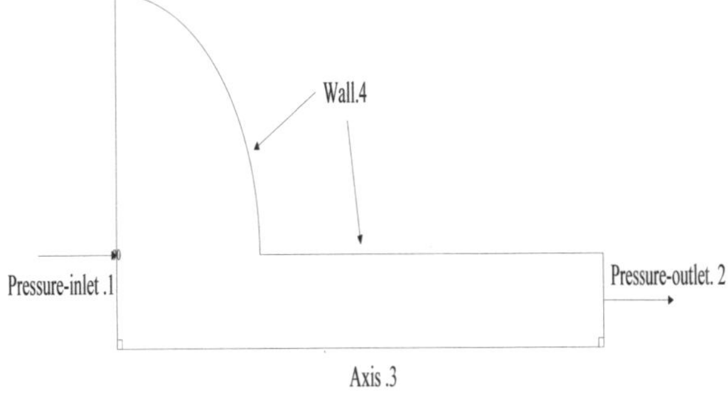

Figure 4.4. *Boundary conditions for nozzles*

4.3.2.1. *Effect of injection pressure on cavitation*

Figure 4.5 shows the internal flow patterns and discharge coefficient of a nozzle when the injection pressure increases. It is transformed from a single phrase flow into cavitation flow when the injection pressure reaches 9 MPa. As the injection pressure continues to increase, vapor develops along the nozzle wall and extends to the exit at 20 MPa, creating "super-cavitation". The discharge coefficient increases with the injection pressure before cavitation starts, but the trend is in the opposite direction after cavitation inception.

The result coincides with the research results measured by Park *et al.* [PAR 08]. Wang researched the influence of fluctuations in injection pressure on cavitation inside a nozzle hole under diesel engine conditions [WAN 08]. The result was that the pressure fluctuation had an obvious influence on the inception of cavitation and quasi-periodic behaviors of super-cavitation.

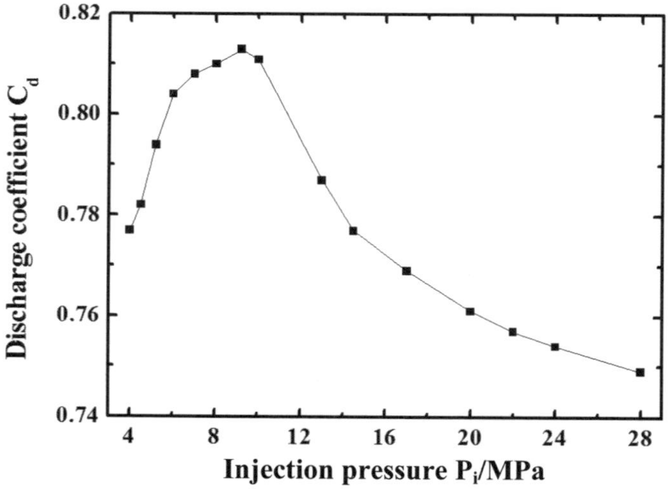

Figure 4.5. *Discharge coefficient of a nozzle with change in injection pressure*

4.3.2.2. *Effect of back pressure on cavitation*

Table 4.1 and Figure 4.6 show the discharge coefficient (C_d), initial amplitude parameter (*amp0*) and distribution of vapor volume fraction of an outlet cross-section at various back pressures when the injection pressure is 30 MPa. With the increase in back pressure, the C_d and *amp0* tend to remain the same. This means that the actual mass flow rate depends on injection pressure alone. It is also consistent with the one-dimensional model proposed by Nurick [NUR 76].

Back pressure P_b/MPa	Actual mass flow rate/g s^{-1}	Discharge coefficient C_d	Initial amplitude parameter *amp0*
0.2	25.95	0.720	33.20
2.0	25.95	0.743	28.98
4.0	25.95	0.771	23.23

Table 4.1. *Comparison of Cd and amp0 at different back pressures*

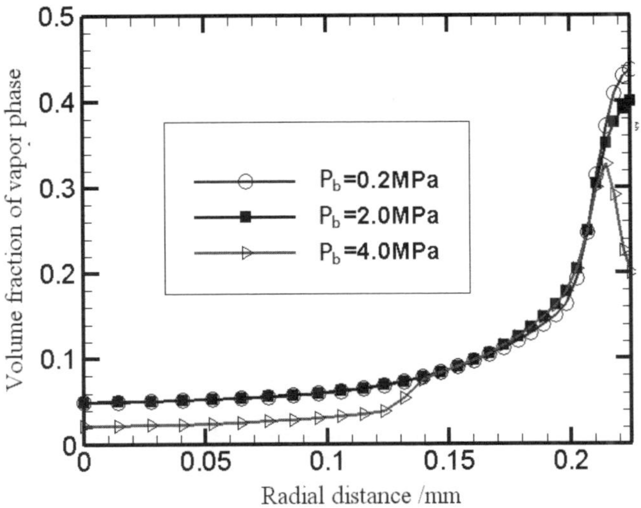

Figure 4.6. *Comparison of distribution of vapor phase fraction at the nozzle outlet under different back pressures*

4.3.3. *Effect of nozzle geometry on cavitation*

4.3.3.1. *Different inlet radii*

Table 4.2 and Figure 4.7 show the discharge coefficient C_d, *amp0* and distribution of vapor volume fraction of an outlet cross-section of nozzles with different inlet radii. With an increase in inlet radius, the discharge coefficient increases, but both *amp0* and vapor volume fraction decrease. Schmidt also found that a nozzle with a rounding inlet produced a much thinner cavitation film [SCH 97b].

Inlet radius r/μm	**Actual mass flow rate/g s^{-1}**	**Discharge coefficient C_d**	**Initial amplitude parameter *amp0***
0	23.81	0.682	36.80
30	25.95	0.743	28.98
60	28.56	0.818	11.19

Table 4.2. *Comparison of the C_d and amp0 of nozzles with different inlet corner radii*

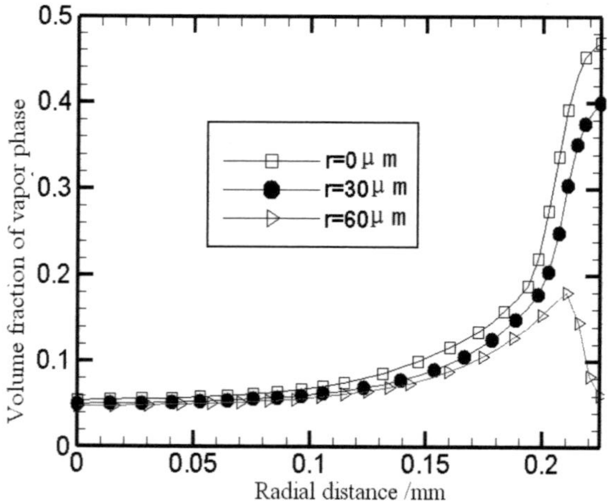

Figure 4.7. *Comparison of the distribution of vapor phase fraction of nozzles with different inlet corner radii*

4.3.3.2. *Different shapes*

Table 4.3 and Figure 4.8 show the discharge coefficients C_d, *amp0* and distribution of vapor volume fractions of the outlet cross-section of nozzles of different shapes. The discharge coefficient of divergent nozzle (K = -0.5) is lower than that of contracting nozzle (K = 0.5). However, the cavitation intensity and *amp0* of the divergent nozzle are higher than those of a contracting nozzle under the same injection conditions. Thus, a divergent nozzle tends to have a better performance. This fact was validated by Payri's experimental studies [PAY 04, PAY 06].

K	Actual mass flow rate/g s^{-1}	Discharge coefficient C_d	Initial amplitude parameter *amp0*
-0.5	24.33	0.697	35.25
0	25.95	0.743	28.98
0.5	28.86	0.826	6.71

Table 4.3. *Comparison of C_d and amp0 of different shaped nozzles*

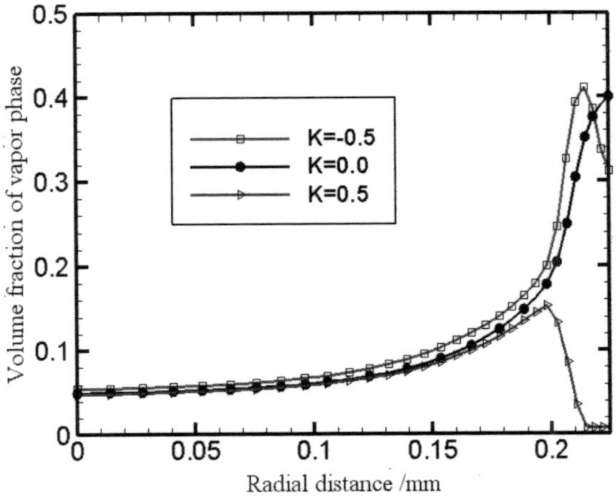

Figure 4.8. *Comparison of distribution of vapor phrase fraction of nozzles with different shapes*

4.3.3.3. *Different length-to-diameter ratios*

As shown in Table 4.4, the discharge coefficient is virtually constant while the L/D ratio changes from 2 to 6. According to Figure 4.9, the distributions in vapor volume fraction of the outlet cross-section of orifices with different L/D ratios vary markedly [WAN 11].

With the increasing L/D ratio, the vapor phase inclines towards the center part of the nozzle and the *amp0* decreases. Thus, we will get a worse spray performance for a nozzle with a bigger L/D ratio.

Schmidt [SCH 97b] also found that longer nozzles tended to show the effect of momentum diffusion in the exit profile without significant change in the coefficient of discharge.

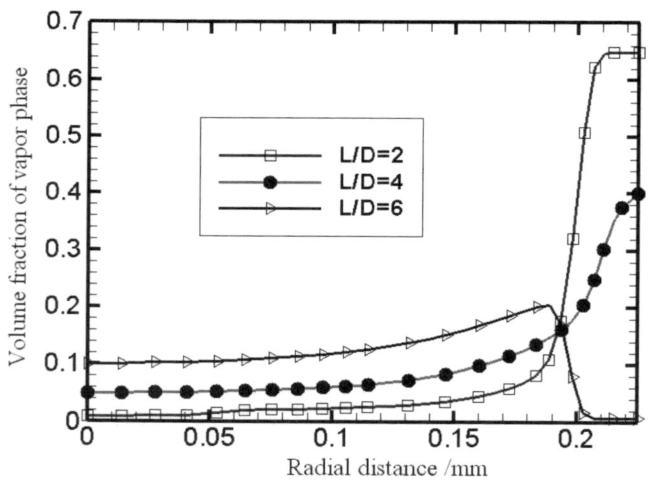

Figure 4.9. *Comparison of distribution of vapor phase fraction of nozzles with different L/D ratios*

Length-to-diameter ratio L/D	Actual mass flow rate/g s^{-1}	Discharge coefficientC_d	Initial amplitude parameter *amp0*
2	26.09	0.747	40.08
4	25.95	0.743	28.98
6	26.09	0.747	23.14

Table 4.4. *Comparison of C_d and amp0 of nozzles with different L/D ratios*

4.4. Influence of fuel property on cavitation.

From the viewpoint of the environment and energy situation [PAR 08], the application of biofuel is under consideration because it can be used without modification of the fuel-supply system in a diesel engine. Furthermore, it can be expected to improve the emission characteristics of such engines. The majority of older studies, however, do not provide an understanding of the formation and influence of cavitation. Most such studies have used water or diesel as the working fluid, which does not reflect the properties of fuel in an internal combustion (IC) engine. From this point of view, studying the effect of fuel properties, such as viscosity and vapor pressure, on cavitation is very necessary [FRA 08].

The difference in cavitating flow between micro-emulsified biofuel and diesel under similar conditions has been compared. Following this comparison, the influence of fuel properties on cavitation has also be analyzed. Micro-emulsificated biofuel was prepared by our research team [XU 09]. Table 4.5 shows the properties of the obtained micro-emulsificated biofuel and diesel it was compared to.

Item	Micro-emulsificated biofuel	0# Diesel
Density (kg m^{-3})	861.3	839.2
Dynamic viscosity (Pa·s)	0.0044	0.003
Surface tension (N m^{-1})	0.045	0.028
Saturated vapor pressure (Pa)	150	300

Table 4.5. *The properties of two fuels (at 300 K)*

4.4.1. *Cavitating flow characteristics*

Table 4.6 provides a comparison of the cavitating flow between micro-emulsificated biofuel and diesel in nozzles with variable injection pressures. As shown in Tables 4.6a and 4.6b, the pressure of inception cavitation of micro-emulsificated biofuel is 8.2 MPa and the mass flow rate is 13.31 $g.s^{-1}$. The diesel has a inception cavitation pressure of 7.4 MPa and mass flow rate of 11.95 $g.s^{-1}$. The injection pressure of transition cavitation and super cavitation of the micro-emulsificated biofuel are 9.2 MPa and 24 MPa, respectively, while those of the diesel are 8.1 MPa and 22.5 MPa, respectively. Thus, we can drawn the conclusion that micro-emulsificated biofuel is hard to cavitate compared to diesel. This result is in in agreement with the experimental results obtained by Suh *et al.* [SUH 08b].

	Single phase flow	Inception cavitation	Development of cavitation		
			Sub-cavitation	Transition cavitation	Super-cavitation
Injection pressure P_1 (MPa)	6.0	8.2	9.0	9.2	24.0
Cavitation number K	1.50	1.32	1.29	1.27	1.09
Mass flow rate (g s^{-1})	9.54	13.31	14.18	14.53	23.23
Flow character-istics inside nozzles		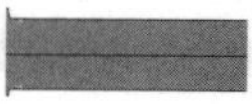			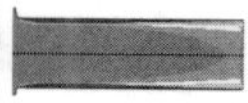

(a) Micro-emulsificated biofuel

	Single phase flow	Inception cavitation	Development of cavitation		
			Sub-cavitation	Transition cavitation	Super-cavitation
Injection pressure P_1 (MPa)	6.0	7.4	7.8	8.1	22.5
Cavitation number K	1.50	1.37	1.34	1.33	1.10
Mass flow rate (g s^{-1})	10.74	11.95	12.51	13.33	22.20
Flow characteri-stics inside nozzles	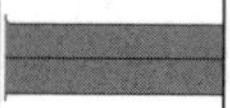		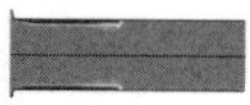	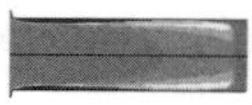	

(b) Diesel

Volume fraction of vapor phase

Table 4.6. *Comparison of the cavitating flow characteristics caused by two fuels inside nozzles*

4.4.2. *Variation in the characteristics of dimensionless parameters*

Figure 4.10 shows the variations in cavitation number, K, and Reynolds number, R_e, against injection pressure. The cavitation numbers of two fuels under a similar injection pressure are considered to be the same, since the pressure is $P_v << P_1$. The inception cavitation number, transition cavitation number and supercavitation number of the micro-emulsificated biofuel are 1.32, 1.27 and 1.09, respectively, which are all higher than those of diesel. Thus, it can be concluded that diesel is liable to cavitate in nozzles. It can also be seen that the Reynolds number of diesel is higher than that of the micro-emulsificated biofuel due to its higher viscosity. This phenomenon reveals that the turbulent effect inside nozzles is greater for diesel.

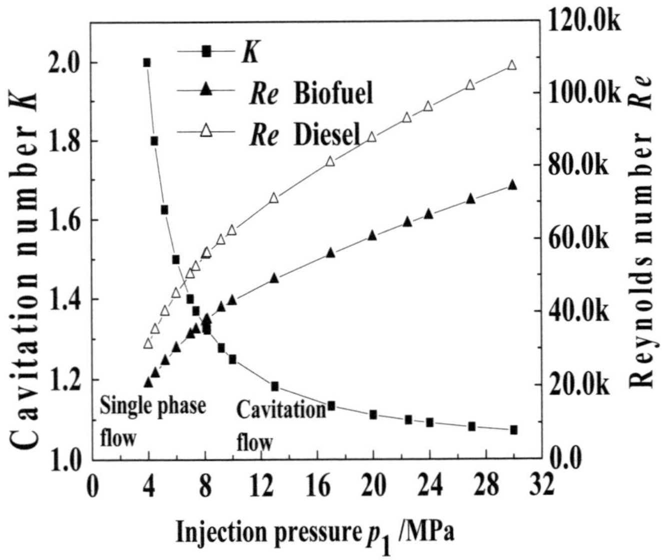

Figure 4.10. *Comparison of the cavitation number and Reynolds number of biofuels and diesel*

Figure 4.11 shows the variations in discharge coefficients due to the injection pressure of two fuels in nozzles. It can be seen that the discharge coefficient increases with injection before cavitation occurs, and that the discharge coefficient of the micro-emulsified biofuel is lower than that of the diesel. When cavitation occurs, the discharge coefficients of two fuels increase continuously and reach their maximum values of 0.815 and 0.827, respectively, at the point of transition cavitation. After transition cavitation has been reached, they both decrease as the rate of injection pressure increases and have a similar tendency.

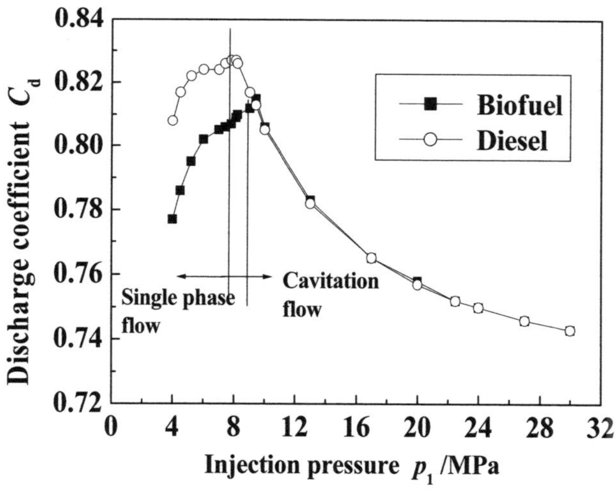

Figure 4.11. *Relationship between the discharge coefficient and injection pressure of biofuel and diesel*

Figure 4.12 shows the variation in initial *amp0* with cavitation number. It can be seen that there is a distinct difference in the initial *amp0* before and after cavitation occurs.

In cases of single phase flow at low pressure, the *amp0* slightly increases with the decrease in cavitation number and the *amp0* of micro-emulsified biofuel is higher than that of diesel. After transition cavitation, they both have a significant increase in *amp0* with decrease in cavitation number. With the development of cavitation in nozzles (cavitation number $K \leq 1.29$), the *amp0* of the micro-emulsificated biofuel is slightly lower than that of diesel. As a result, we find that the micro-emulsificated biofuel has a worse spray performance inside the high-pressure nozzle with cavitation compared with diesel.

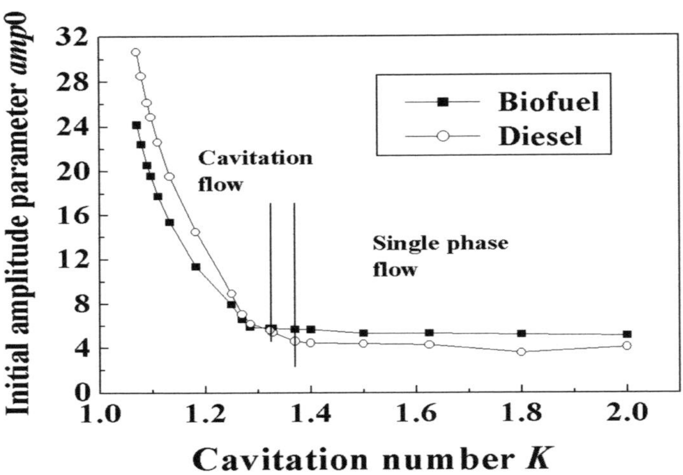

Figure 4.12. *Relationship between the initial amp0 and the cavitation number*

4.4.3. *Effect of fuel properties on cavitation inception*

The effect of the characteristics of fluids on cavitation inception in an orifice has also been investigated by us. The numerical simulation cavitation model that was section 4.3.1 was implemented when the fluids had different properties (viscosity, vapor pressure and surface tension).

Fluid	**1**	**2**	**3**	**4**	**5**	**6**	**7**	**8**	**9**
Dynamic viscosity (Pa.s)	0.001	0.001	0.001	0.005	0.005	0.005	0.01	0.01	0.01
Surface tension (N m^{-1})	0.02	0.02	0.02	0.02	0.02	0.02	0.02	0.02	0.02
Saturated vapor pressure (Pa)	300	3000	6000	300	3000	6000	300	3000	6000
Fluid	**10**	**11**	**12**	**13**	**14**	**15**	**16**	**17**	**18**
Dynamic viscosity (Pa.s)	0.001	0.001	0.001	0.005	0.005	0.005	0.01	0.01	0.01
Surface tension (N m^{-1})	0.07	0.07	0.07	0.07	0.07	0.07	0.07	0.07	0.07
Saturated vapor pressure (Pa)	300	3,000	6,000	300	3,000	6,000	300	3,000	6,000

Table 4.7. *Characteristics of 18 kinds of fluids*

Eighteen kinds of fluids with different properties were compared. The characteristics of these are shown in Table 4.7.

The formation and development of cavitation inside the orifice was quantitatively analyzed using parameters such as critical cavitation pressure and critical cavitation number. According to numerical simulation, it was revealed that the critical cavitation pressure increased with the increase of fluid viscosity and decrease in the vapor pressure of the fluid. The surface tension of the fluid, however, was not related to the cavitation inception. This result was also in agreement with the experimental results published by of Suh [SUH 08b] and Franzoni [FRA 08].

4.5. Cavitation erosion of biofuel in the diesel injection nozzle

4.5.1. *Effect of cavitation erosion on a nozzle*

Cavitation in a diesel injection nozzle can cause a degradation in engine performance and structural damage. Asi studied a damaged diesel engine injector nozzle, which was made from 18CrNi8 steel [ASI 06]. This nozzle was found to be cracked after about 400 hours in service. The appearance of this crack can be seen Figure 4.13.

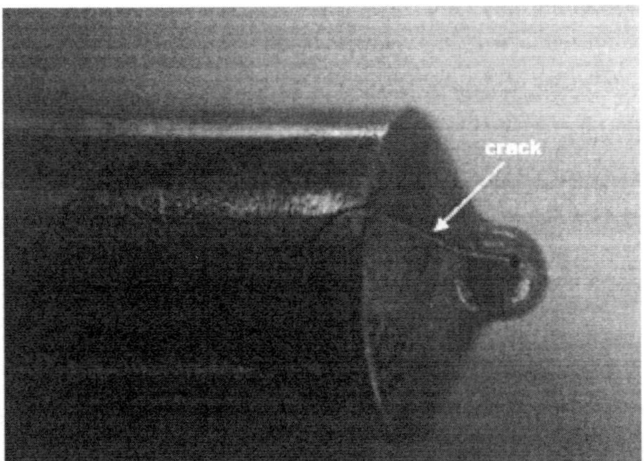

Figure 4.13. *The appearance of cracks*

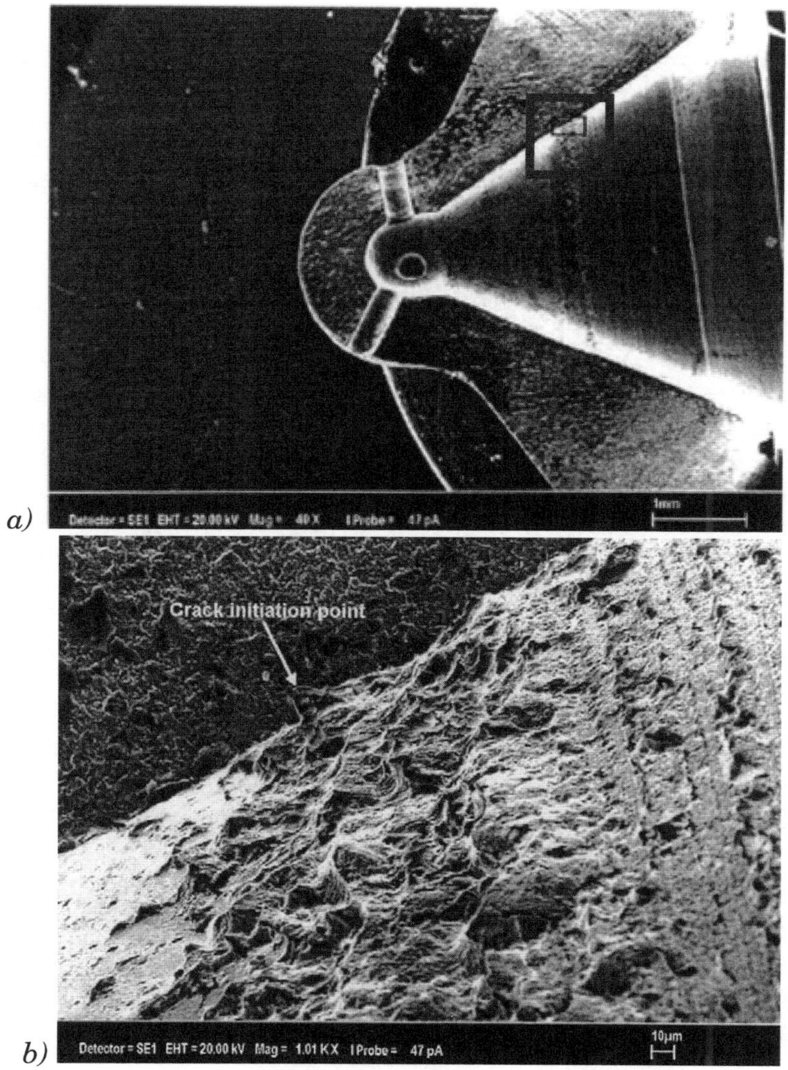

Figure 4.14. *Failed nozzle after separation: (a) the typical appearance of a macro-fracture; (b) Magnified view of the box outlined in black in Figure 4.14a*

The cross-section surface of the failed nozzle after being cut is shown in Figure 4.14a. Examination of the surface showed it to be extensively damaged at some locations.

The damaged surfaces had common features including extensive plastic deformation of various regions combined with metal removal, the formation of pits and surface depressions. These features have been associated with cavitation damage.

Figure 4.14b shows the cavitation-damaged areas indicated by the box in Figure 4.14a at a higher magnification. Localized plastic deformation and crack-initiation sites are clearly visible on the cavitation-damaged surface. This scanning electron micrograph (SEM) revealed that crack initiation occurred at the cavitation-damaged areas and propagated on the subsurface as a fatigue crack.

4.5.2. *Location of cavitation erosion in a nozzle*

Gavaises [GAV 07] used enlarged transparent replicas of nozzles to investigate the formation and development of cavitation inside diesel injector nozzles suffering from erosion damage. The transparent nozzles were manufactured from acrylic. The two nozzle geometries that have been found to suffer from cavitation erosion are shown in Figures 4.15 and 4.16.

The first nozzle (Figure 4.15a) was a value cover orifice (VCO). For this nozzle, two sites of surface erosion have been identified (see Figures 4.15b and 4.15c). The first is found at the top part of the hole, close to the entry of the fuel. The second erosion site has been observed at the 3 and 9 o'clock positions of the entrance to the hole, as shown in Figure 4.15c.

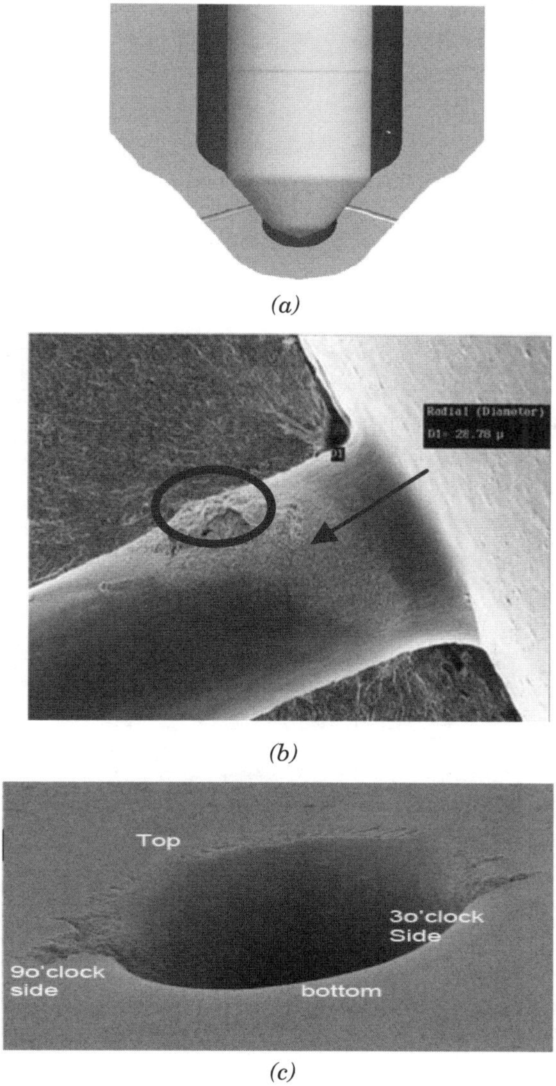

Figure 4.15. *(a) Schematic of the VCO nozzle; (b) picture of erosion damage at the top of the injection hole; (c) erosion damage at the two sides of injection hole*

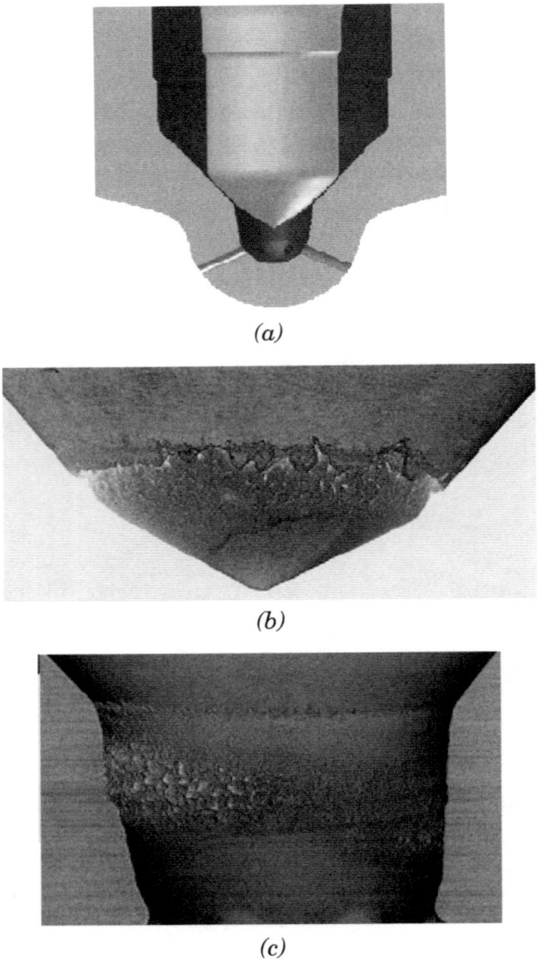

Figure 4.16. *(a) Schematic of the sac-type nozzle; (b) picture of erosion damage of the needle; (c) erosion damage of the sac volume*

The second nozzle tested was a sac-type nozzle. For this nozzle, cavitation erosion has been observed at the needle itself (see Figure 4.16b) as well on the sac-type volume surface upstream of the hole inlet (Figure 4.16c). The holes themselves have suffered much less erosion compared to the VCO nozzle.

4.5.3. *Factors that influence cavitation erosion in nozzles*

4.5.3.1. *Hardness of surface*

In order to investigate the effect of the hardness of the needle-sealing surface on cavitation erosion, two kinds of nozzles were selected and operated in a S195 diesel engine for 10 hours. One nozzle was heat-treated with a micro-hardness of 420 HV; the other had a micro-hardness of 840 HV and is a commercial-used nozzle. The appearances of the surfaces of the original and operated needle-seal were observed using a SEM (see Figure 4.17).

It was found that the cavitation erosion on the seal surface of the nozzle with low hardness after the annealing process was worse than that of the normal nozzle. Their patterns of wear and formations, however, are similar.

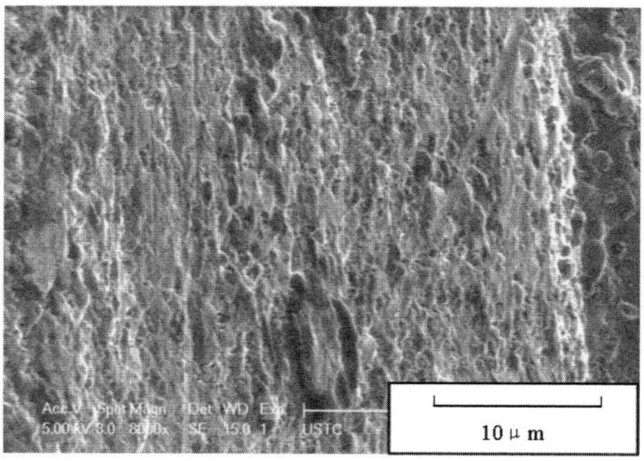

(a)

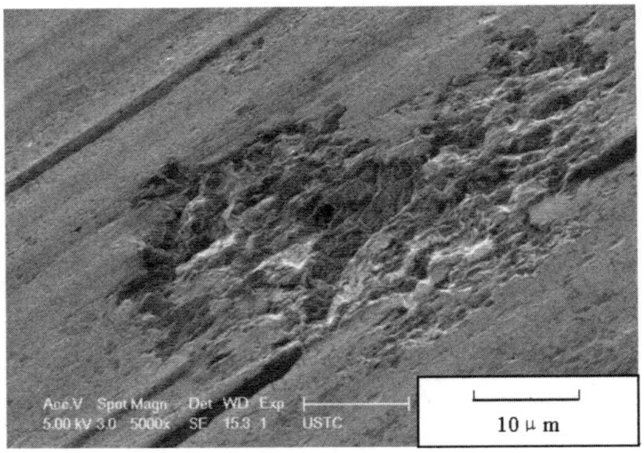

(b)

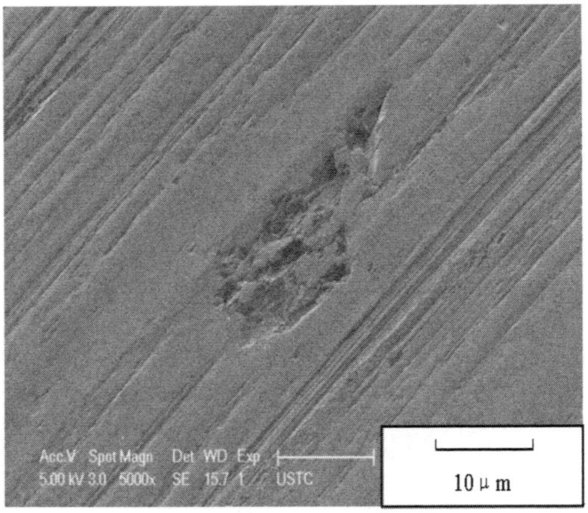

(c)

Figure 4.17. *SEM images of a needle-sealing surface: (a) original state of the seal surface; (b) seal surface of the heat-treated needle with that was operated for 10 hours; (c) seal surface of the commercial needle which was operated for 10 hours*

4.5.3.2. *Original surface defects*

The original defects on a surface have an effect on cavitation erosion. The appearance of the original needle-sealing surface is shown in Figure 4.18.

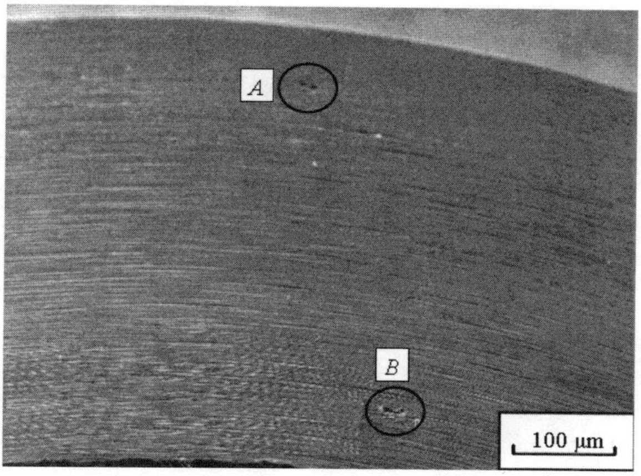

Figure 4.18. *SEM image of the original needle-sealing surface*

In Figure 4.18 we can see two original surface defects, marked A and B. A is located upstream and B is located downstream. After the needle was operated for 20 hours in a S195 diesel engine, the surface appearances of the operated needle-sealing surface were observed by SEM (see Figure 4.19).

The results show that the original pits located upstream have obvious wear after the diesel engine had operated for 20 hours. However, this phenomenon cannot be shown in the downstream direction. The location of severe wear is in the same place as the severe cavitation area. The reason for the wear is that the original defects (pits) tend to collect the wandering cavitation bubbles [BAI 09]. The wear on the solid wall is the result of the collapse of cavitation bubbles.

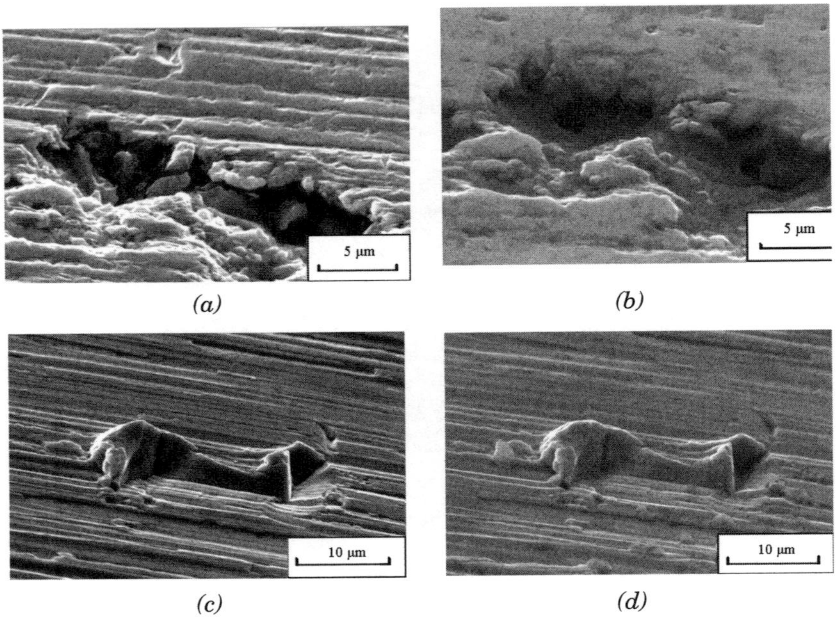

Figure 4.19. *SEM images of defects on the original and operated needle-sealing surface: (a) defect A on the original needle-sealing surface; (b) defect of A on the operated needle-sealing surface; (c) defect of B on the original needle-sealing surface; (d) defect of B on the operated needle-sealing surface*

4.5.4. *Effect of biofuel on the erosion of nozzles*

Chiaramonti researched the effects of various BCO (bio crude oil)/diesel emulsions on the injection systems of different diesel engines [CHI 03]. 18CrNi9 sheet metal (similar to the nozzle material, 18CrNi8) was wetted with 75% emulsion for three days, after which the surface showed no visible traces of corrosion. Emulsions do not appear to be corrosive to the material nozzles are made from.

Nozzles are, however, seriously damaged and the injection channels greatly enlarged when the diesel engine is fed with emulsions. The same marks were found on the pump elements as we found on the needles, but – as expected – over a longer period of time.

It is therefore possible that damage to the nozzles is caused by a complex mechanism that combines corrosion with the effects of the high-velocity, high-pressure turbulent flows occurring in the injection channels, such as cavitation. In fact, as already explained in the literature, cavitation is the dominant mechanism responsible for long-term damage of diesel injection nozzles, even under normal operating conditions when fossil fuels that are neither corrosive nor contaminated by particles are used.

4.6. Conclusion

There are a lot of bubbles in a nozzle before the needle opens. The cavitation located on upstream of the seal surface is greater than that downstream. In the process of bubbles collapsing, a cavity is thought to be responsible for the damage caused by cavitation. The cavitation and wear characteristics of the seal surface in the nozzle have important influences on the atomization, combustion and dynamic performance of diesel engines.

A homogeneous mixture flow model with a full cavitation model can accurately forecast the cavitation flow inside an injector nozzle. Increasing the injection pressure or decreasing back pressure can enhance the cavitation effect. The increasing inlet corner radius may weaken cavitation. The role of cavitation in the divergent nozzle is higher than that in the convergent nozzle. Decreasing the L/D ratio can improve the spray ability, as the flowing coefficient stays the same. Higher cavitation leads to a better spraying effect and to a lower flow coefficient.

The characteristics of fluids have effects on cavitating flow. Due to the different characteristics of micro-emulsificated biofuel and diesel, we needed to study different injection pressures at different cavitation stages, including inception cavitation, transition cavitation and super cavitation. Cavitating flow characteristics include such aspects as:

– the mass flow rate of the nozzle outlet;

– the dimensionless parameter the Reynolds Number (R_e);

– the initial (*amp0*);

– the discharge coefficient (C_d).

All of these were different under the same working conditions. Critical cavitation pressure increased with the increase in fluid viscosity and decrease in vapor pressure of the fluid. The surface tension of the fluid, however, was not relative to cavitation inception.

Cavitation in diesel injection nozzles can cause structural damage to the engine. Cracks originating from cavitation-damaged areas can propagate on the subsurface as fatigue cracks. For nozzles with a different geometry, the locations of cavitation damage are different. The location of severe wear is the same place as the area of severe cavitation. Cavitation erosion on the surface of the softer nozzle was worse than that of the harder nozzle. Original defects (pits) on the surface of a nozzle tend to collect the wandering cavitation bubbles and cause more serious damage. In the case of biofuel, it is possible that damage to the nozzles is caused by a complex mechanism that combines corrosion with the effects of cavitation.

4.7. Acknowledgments

The authors want to express their thanks to Professor Xifeng Zhu from the University of Science and Technology of China, and Mr. Guofeng Wang and Mr Wendong Li from

Hefei University of Technology for their direction and assistance in the computation, experiment and discussion, respectively. Financial support from the National Natural Science Foundation of China (Grant No. 50875071) is also acknowledged. This work was also supported by a grant from the National High Technology Research and Development Program of China (863 Program, Grant No. 2012AA051803).

4.8. Bibliography

[ASI 06] ASI O., "Failure of a diesel engine injector nozzle by cavitation damage", *Engineering Failure Analysis*, vol. 13, p. 1126-1133, 2006.

[BAD 99] BADOCK C., WIRTH R., FATH A., LEIPERTZ A., "Investigation of cavitation in real size diesel injection nozzles", *International Journal of Heat & Fluid Flow*, vol. 20, p. 538-544, 1999.

[BAI 09] BAI L. X., XU W.L., ZHANG F.X., LI N.W., ZHANG Y.C., HUANG D.F., "Cavitation characteristics of pit structure in ultrasonic field", *Science in China* (Series E), vol. 52, no. 7, p. 1974-1980, 2009.

[BEN 66] BENJAMIN T.B., ELLIS A.T., "The collapse of cavitation bubbles and the pressure thereby produced against solid boundaries", *Philosophical Transactions*, vol. 260, no. 1110, p. 221-240, 1966.

[BRE 05] BRENNEN C.E., *Fundamentals of Multiphase Flow*, Cambridge University Press, New York, USA, 2005.

[CHA 95] CHAVES H., KNAPP M., KUNITZEK A., OBERMEIER F., SCHNEIDER T., "Experimental study of cavitation in the nozzle hole of diesel injectors using transparent nozzles", *SAE Paper 950290*, 1995.

[CHE 02] CHEN W.S., LAFON C., MATULA T.J., VAEZY S., CRUM L.A., "Mechanisms of lesion formation in high intensity focused ultrasound therapy", *2002 IEEE Ultrasonics Symposium*, Munich, Germany, vol. 2, p. 1443-1446, 2002.

[CHE 06] CHEN H., LI X.J., WAN M.X., "The inception of cavitation bubble clouds induced by high-intensity focused ultrasound", *Ultrasonics*, vol. 44, no. 1, p. 427-429, 2006.

[CHE 08a] CHEN H.S., LI J., "A ring area formed around the erosion pit on 1Cr18Ni9Ti stainless steel surface in incipient cavitation erosion", *Wear*, vol. 266, p. 884-887, 2008.

[CHE 08b] CHEN H.S., LIU S.H., "Inelastic damages by stress wave on steel surface at the incubation stage of vibration cavitation erosion", *Wear*, vol. 266, p. 69-75, 2008.

[CHI 03] CHIARAMONTI D., BONINI M. FRATINI E., "Development of emulsions from biomass pyrolysis liquid and diesel and their use in engines – Part 2: tests in diesel engines", *Biomass and Bioenergy*, vol. 25, p. 101-111, 2003.

[FRA 08] FRANZONI F, MILANI M, MONTORSI L. "The influence of cavitation and aeration in a multi-fuel injector", *SAE Paper 2008-01-2390*, 2008.

[GAN 01] GANIPPA L.C., BARK G., ANDERSSON S., CHOMIAK J., "Comparison of cavitation phenomena in transparent scaled-up single-hole diesel nozzles", CAV 2001: *4th International Symposium on Cavitation*, California Institute of Technology, Pasadena, CA, USA, p. 1-9, June 20-23, 2001.

[GAV 60] GAVRANEK V.V., BOLSHUTKIN D.N., ZELDOVICH V.I., "Thermal and mechanical action of a cavitation zone on the surface of a metal", *Fiz. Metal I. Metalloved*, vol. 10, no. 2, p. 262-268, 1960.

[GAV 07] GAVAISES M., PAPOULIAS D., ANDRIOTIS A., GIANNADAKIS E., "Link between cavitation development and erosion damage in diesel injector nozzles", *SAE Paper 2007-01-0246*, 2007.

[GIA 04] GIANNADAKIS E., GAVAISES M., ROTH H., ARCOUMANIS C., "Cavitation modelling in single-hole diesel injector based on Eulerian-Lagrangian approach", *THIESEL 2004 Conference on Thermo and Fluid Dynamic Processes in Diesel Engines*, Valencia, Spain, p. 1-13, September 11-13, 2004.

[GIA 07] GIANNADAKIS E., PAPOULIAS D., GAVAISES M., ARCOUMANIS C., SOTERIOU C., TANG W., "Evaluation of the predictive capability of diesel nozzle cavitation models", *SAE Paper 2007-01-0245*, 2007.

[HAM 80] HAMMITT F.G., *Cavitation and Multiphase Flow Phenomena*, McGraw-Hill, London & New York, 1980.

[HU 11] HU X.G., WANG G.F., WO H.Z., XU Y.F., "Cavitation flow properties of micro-emulsified biomass fuel in injector nozzle", *Journal of Mechanical Engineering*, vol. 47, no. 12, p. 99-105, 2011.

[LIU 08] LIU J.S., FU Z., XU C.S., "Research on visualization of cavitation in spraying of diesel engine", *Chinese Internal Combustion Engine Engineering*, vol. 31, no. 4, p. 89-95, 2008.

[LIV 08] LIVERANI L., ARCOUMANIS C., YANAGIHARA H., SAKATA I., OMAE K., "Imaging of the flow and cavitation formation in a transparent real-size six-hole nozzle under realistic conditions", *COMODIA*, Hokkaido University, Japan, p. SC14, July 28-31, 2008.

[MOM 95] MOMMA T., LICHTAROWICZ A., "A study of pressures and erosion produced by collapsing cavitation", *Wear*, vol. 186-187, p. 425-436, 1995.

[NOW 42] NOWOTNY H., *Destruction of Materials by Cavitation*, VDI-Verlag, Berlin, Germany, 1942.

[NUR 76] NURICK W.H., "Orifice cavitation and its effects on spray mixing", *Journal of Fluids Engineering*, vol. 98, p. 681-687, 1976.

[OKA 94] OKADA T., IWAI Y., ISHIMARU H., MAEKAWA N., "Measurement and evaluation of cavitation bubble collapse pressures", *JSME International Journal Series A, Mechanics and Material Engineering*, vol. 37, no. 1, p. 37-42, 1994.

[PAR 08] PARK S.H., SUH H.K., LEE C.S., "Effect of cavitating flow on the flow and fuel atomization characteristics of biodiesel and diesel fuels", *Energy & Fuels*, vol. 22, p. 605-613, 2008.

[PAY 04] PAYRI F., BERMUDEZ V., PAYRI R., SALVADOR F.J., "The influence of cavitation on the internal flow and the spray characteristics in diesel injection nozzles", *Fuel*, vol. 83, p. 419-431, 2004.

[PAY 06] PAYRI F., ARRÈGLE J., LÓPEZ J.J., HERMENS S., "Effect of cavitation on the nozzle outlet flow, spray and flame formation in a diesel engine", *SAE paper 2006-01-1391*, 2006.

[PLE 71] PLESSET M.S., CHAPMAN R.B., "Collapse of an initially spherical vapour cavity in the neighbourhood of a solid boundary", *Journal of Fluid Mechanics*, vol. 47, p. 283-290, 1971.

[PLE 77] PLESSET M.S., PROSPERETTI A., "Bubble dynamics and cavitation", *Annual Review of Fluid Mechanics*, vol. 9, p. 145-185, 1977.

[RAY 17] RAYLEIGH L., "On the pressure developed in a liquid during the collapse of a spherical cavity", *Philosophical Magazine*, vol. 34, no. 200, p. 94-98, 1917.

[SAT 01] SATO K., SAITO Y., "Unstable cavitation behavior in circular-cylindrical orifice flow", *JSME International Journal Series B: Fluids and Thermal Engineering*, vol. 45, no. 3, p. 638-645, 2002.

[SCH 97a] SCHMIDT D.P., CORRADINI M.L., "Analytical prediction of the exit flow of cavitating orifices", *Atomization and Sprays*, vol. 7, no. 6, p. 54-65, 1997.

[SCH 97b] SCHMIDT D.P., RUTLAND C.J., CORRADINI M.L., "A numerical study of cavitating flow through various nozzle shapes", *SAE Paper 971597*, 1997.

[SCH 99] SCHMIDT D.P., RUTLAND C.J., CORRADINI M.L., ROOSEN P., GENGE O., "Cavitation in two-dimensional asymmetric nozzles", *SAE Paper 1999-01-0518*, 1999.

[SCH 08] SCHAUB G., VETTER A., "Biofuels for automobiles – an overview", *Chemical Engineering & Technology*, vol. 31, no. 5, p. 721-729, 2008.

[SIN 02] SINGHAL A.K., ATHAVALE M.M., LI H.Y., JIANG Y., "Mathematical basis and validation of the full cavitation model", *Journal of Fluids Engineering*, vol. 124, no. 3, p. 617-624, 2002.

[STE 02] STEPHANE P., STEPHANE Z., "Bubble collapse near a solid boundary: A numerical study of the influence of viscosity", *Journal of Fluid Mechanics*, vol. 464, no. 8, p. 137-163, 2002.

[SUH 08a] SUH H.K., LEE C.S., "Effect of cavitation in nozzle orifice on the diesel fuel atomization characteristics", *International Journal of Heat and Fluid Flow*, vol. 29, p. 1001-1009, 2008.

[SUH 08b] SUH H.K., PARK S.H., LEE C.S., "Experimental investigation of nozzle cavitating flow characteristics for diesel and biodiesel fuels", *International Journal of Automotive Technology*, vol. 9, no. 2, p. 217-224, 2008.

[TAK 05] TAKENAKA N., KADOWAKI T., KAWABATA Y., LIM I.C., SIM C.M., "Visualization of cavitation phenomena in a Diesel engine fuel injection nozzle by neutron radiography", *Nuclear Instruments and Methods in Physics Research Section A: Accelerators, Spectrometers, Detectors and Associated Equipment*, vol. 542, no. 1-3, p. 129–133, 2005.

[WAN 08] WANG X., SU W., "Influence of injection pressure fluctuations on cavitation inside a nozzle hole at diesel engine conditions", *SAE Paper 2008-01-0935*, 2008.

[WAN 11] WANG G.F., WO H.Z., ZHANG Y.F., XU Y.F., HU X.G., "Study of cavitation flow performance in diesel injector nozzle", *6th China International Symposium on Tribology*, Lanzhou, China, August 19-22, 2011.

[WAT 96] WATKIN N.A., HAAR G.R., RIVENS I., "The intensity dependence of the site of maximal energy deposition in focused ultrasound surgery", *Ultrasound in Medicine & Biology*, vol. 22, no. 4, p. 483-491, 1996.

[WU 93] WU C.C., ROBERT P.H., "Shock-wave propagation in a sonoluminescing gas bubble", *Physics Review Letters*, vol. 70, no. 5, p. 3424-3427, 1993.

[XU 09] XU Y.F., WANG Q.J., HU X.G., ZHU X.F., "Preparation and tribological performance of micro-emulsified bio-oil", *Acta Petrolei Sinica* (Petroleum Processing Section), vol. 25, no. S1, p. 53-56, 2009.

[YIN 02] YING C.F., AN L., "High temperature and high pressure distribution in gas bubbles generated by sound cavitation", *Science in China* (Series A), vol. 32, p. 305-313, 2002.

[ZHO 83] ZHOU Y.K., HAMMITT F.G., "Cavitation erosion incubation period", *Wear*, vol. 86, no. 2, p. 299-313, 1983.

Chapter 5

Wear and Corrosion Damage of Medical-grade Metals and Alloys

Damage and failure of metallic implant surfaces may occur due to synergistic interactions between repeated contact loading and corrosion. A number of factors determine the rate and progression of surface damage including:

– the mechanical properties of mating surfaces;

– properties of the passivating oxide layers;

– contact loading;

– relative motion;

– residual stresses;

– surface roughness;

– chemical environment.

The influence of various factors and recent research results on the interrelationship between surface roughness,

---

Chapter written by Jae-Joong RYU and Pranav SHROTRIYA.

mechanical loading and corrosive environment are discussed in this chapter.

5.1. Introduction

The most common cause of failure in orthopedic implants is a multi-factorial phenomenon that includes fatigue, wear, fracture, corrosion and infection.

Implant failure usually includes loosening or breakage of the internal fixation device. Since hard tissue (bone structure) is more flexible than metals, at the fixation with screws, stress concentration will be produced on the implant materials that leads to breakage and misalignment of the stem inserted into the bone. Fatigue loading on the implants also initiates and propagates cracks. After the steady growth of a crack, it reaches a critical length and the stress drastically increases at the crack-tips. Brittle fracture is an uncommon failure process, but it occurs when there is a flaw in the implant design that initially establishes a stress raiser beyond the yield point of the implant material.

Fatigue contact and fretting are the most common causes of mechanical damage at implant surfaces. Repetitive frictional sliding motion on the bearing surface of joint replacements produces wear damage and continuous relative displacement is repeated at modular interfaces during human activities. Cyclic, small-scale, frictional relative motion at the interface between components is known as fretting fatigue. For example, during human activities such as walking, jogging and climbing, on a microscale, relative motion at the taper junction of a total hip joint replacement leads to an oscillating friction force at the head–stem junctions that accumulates localized plastic deformation at the asperities in contact. The long-range plastic and elastic deformation initiates micro-cracks near the surface where the interaction is taking place.

In the following sliding contact loadings, crack-tips undergo high stress concentrations due to shear and opening. High stress applied at a crack-tip will be in the range of the tensile and compressive yield points that ultimately lead to a decrease in the fatigue strength of the implant material. As a result, the cracks coalesce and propagate over subsequent repeated contact – the phenomenon known as delamination wear [JAH 74, SUH 73].

Orthopedic implants are subjected to repetitive relative loadings in a physiological environment. The patient's motions in daily living bring implant components into cyclic contact in a corrosive environment. Early retrieval studies of hard tissue replacements showed mechanical and corrosion degradation of implant surfaces [WIL 76]. The stress concentration by fatigue loading and cyclic relative motion at modular connection induces small-scale plastic deformation and initiates implant fracture [VIC 96, VIC 97]. Corrosion damage occurs simultaneously, due to the electrochemical potential that mainly arises from oxygen contained in the body's solution, created due to foreign body reactions and a mixture of metal alloys.

Currently, the complicated combined effect of mechanical and chemical damage is being investigated. As a result of such research, oxide layer instability and stress-assisted dissolution mechanisms have been introduced to research [HAL 03a, MIT 08, RYU 07].

The particulate wear debris from implants causes inflammatory events as well as potential immunological problems. A specific immune response against the foreign body is associated with an inflammatory process that results in necrosis of bone tissue (tissue death). The foreign body reaction leads to the production of macrophages that attempt to degrade and remove the material. The engulfment is not always successful, however, in completely removing the materials and separating them from the surrounding cells. As the reaction is dependent on the nature of the implant material, the variation in biocompatibility will be determined by the physical and chemical condition of the implant surfaces.

Particular chemical components and surface textures may improve bone tissue growth and bone-implant integration. The host response against implants is a complicated phenomenon, however, because the physiological (biochemical) environment continuously changes the properties of the implant surface as a patient's physical conditions and activities change. Conversely, the biological and chemical constitutions of an implant are affected by the mechanical and electrochemical properties of the materials it is made from. The chemical compositions of the corrosion products of wear debris and oxide film breakage causes aggressive immune responses, such as protein adsorption, macrophage and lymphocyte attack. These cell responses may result in necrosis and osteoblast. Such inflammatory events ultimately lead to the premature failure of implants [HAL 03a, JAC 03].

From a chemical point of view, the reaction at the implant surface can be characterized by a thermodynamic driving force that consists of the chemical potentials across the metal–solution interface [AZI 91, BAR 98] while the host response is related to electrical and mechanical properties, such as surface topography, hydrophobicity, surface energy and surface molecular mobility [TEM 08]. The mechanical loading due to human behaviors is also mutually interdependent on the driving forces that synergistically amplify the surface damage mechanism. The combination of these factors induces the accelerated dissolution and premature failure an implant and ultimately degrades its structural integrity and load bearing capacity [MIT 05, MIT 07].

5.1.1. *Total joint replacements*

Joint implants have been designed to mimic joint motion in the form of a pivot hinge or rotational ball-and-socket assembly. In the example of a modular hip joint replacement, the femoral head and stem are the major components that are required to establish joint motion, as shown in Figure 5.1.

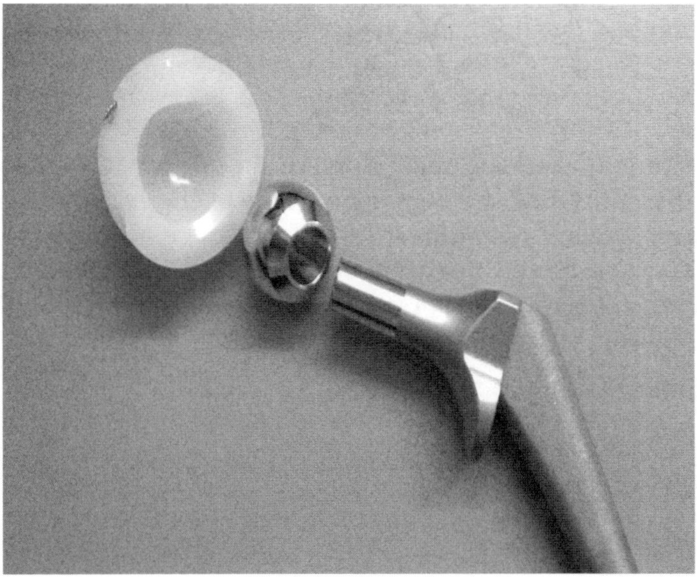

Figure 5.1. *Major femoral component of total hip joint replacement: the polymeric acetabular cup, femoral head and femoral stem*

Early joint replacements used stainless steel and polyethylene inserts for the bearing-socket couple. Cobalt-chromium and titanium alloys have since replaced stainless steel.

The joint replacements made from stainless steel were not successful, due to the poor *in vivo* corrosion resistance of stainless steel led to an adverse effect with the body's cells. Efforts to enhance corrosion resistance led to the development of new materials, such as cobalt-chromium, which achieved greater success. The high frictional forces at the bearing surface, however, resulted in wear and premature loosening of such joints. Polytetraflyororthylene (Teflon) was inserted between the acetabular and femoral components to reduce the friction forces but was not successful due to low wear resistance on the insert. High-density polyethylene then was employed to improve wear resistance and this became basis for the modern design of the total hip joint replacement.

Ceramic materials, such as alumina or zirconia, have also been introduced at the bearing surface of the femoral head to improve wear resistance. For wide customization for patients, assembled modular design has been preferred. For example, total hip joint replacement consists of four major components:

- femoral stem;
- femoral head;
- polymer socket;
- acetabular cup.

In the lower limb connection, the modular implant is constructed by interlocking the conical tapered surfaces between the femoral head and femoral stem. At the tapered connection, the interlocked surface created by machining undergoes plastic deformation by rough surface contact and experiences a cyclic process of oxide film breakage and reformation by the combined effects of contact and corrosion.

The major problem in joint replacements is loosening by wear particles because the hostile tissue response against them and their corrosion products leads to inflammatory events and bone loss. The modular interface deterioration at the self-interlocked connection can be affected by the crevice conditions created by neighboring components. The crevice condition in modular implants limits the access of oxygen to the small space between the two surfaces. As the corrosion takes place, metal oxidation is unstable and the pH level decreases. The pH drop accelerates metal oxidation. The degradation of the protective oxide layer on the surface of the metal implant in crevice conditions cannot securely block the metal surface from solution and prevent further corrosion damage. Accordingly, it is necessary to identify the nature of the surface properties and oxide layer in biomedical materials.

5.1.2. *Metal alloys*

Metal alloys are widely used for broad applications, such as fixing fractures, replacing joints, external splints, braces, traction apparatus and dental amalgams [MUD 03].

In orthopedic implants, the major properties required are a high yield point, good ductility, high fracture toughness, desired corrosion resistance and stable compatibility in the human body. The high yield point and ductility of metals make them suitable for high load-supporting implants without significant dimensional changes due to loadings during implant operations. Another appropriate property is that of fatigue strength and fracture resistance. The toughness of a material is measured by determining the resistance against cyclically repeating loadings. The toughness of metals is superior to ceramics and polymers; however, metals have a relatively poor corrosion resistance in comparison.

As discussed, the wear debris and products due to corrosion cause problems with biocompatibility that it is very important to avoid, as it can lead to an allergic response in the host tissue. Some of metal alloys are suitable, since they satisfy the important properties required. Another reason is the ease with which they can be formed and manufactured into a complex geometry.

The dominant metal alloys used in artificial joints are:

– stainless steel;

– cobalt-based alloy;

– titanium-based alloy.

Cobalt-chromium alloy			
Material (ASTM)	Elastic modulus	Yield strength	Fatigue strength
CoCrMo (F75)	210–253	448–841	207–950
CoCrWNi (F90)	210	448–1,606	586–1,220
CoNiCrMo (F652)	200–230	300–2,000	340–520
Titanium alloy			
CPTi (F67)	110	485	300
Ti6A14V (136)	116	897–1,034	620–689

Table 5.1. *Mechanical properties of Ti6Al4V and CoCrMo*

Their oxide layer, which separates the underlying metal from the reactive environment, is the primary reason for their resistance to corrosion.

The alloys chosen for implants are not perfectly inert in chemical reaction, but the native oxide film blocks the metals from interaction with oxygen and moisture.

Stainless steel alloys were the first metallic materials used for biomedical implants. However their poor resistivity to wear and corrosion limited their service life, while the high ductility was attractive for the dynamic load-bearing use.

Owing to their high-quality fatigue resistance, cobalt-chromium alloys have been selected for joint replacements. The most attractive alloy for implant devices is titanium and its alloys that have low density, good strength and can easily be fabricated.

5.1.2.1. *Stainless steel alloys*

The most common stainless steel alloys used for implants are 316/316L. Other chemical components are added to stainless steel to enhance its corrosion resistance. The composition of stainless steel 316/316L is iron (60–65%), chromium (17–20%), nickel (12–14%) and minor portions of nitrogen, manganese, molybdenum, etc. The significant function of chromium is to establish a corrosion barrier on the surface of the implant. It reacts with oxygen to form a Cr_2O_3 layer which protects the surface from electrochemical reactions in a corrosive environment. This layer is electrochemically stable and prevents the oxidation of stainless steel. Nickel and molybdenum are added to stabilize the structural phase in ferrite and austenite. Molybdenum is also added to improve the corrosion resistance of the grain boundary. 316/316L also contains a small amount of carbon ($<0.03\%$) to suppress corrosion *in vivo*. The work-hardened steels have remarkably higher mechanical strengths, yield strength, tensile strength and fatigue strength, as shown in Table 1.1.

5.1.2.2. *Cobalt–chromium alloys*

Owing to their high fatigue resistance, cobalt–chromium alloys are used in joint replacements. The most prevalent alloys are cobalt–chromium–molybdenum (CoCrMo), cobalt–

nickel–chromium–molybdenum (CoNiCrMo) and cobalt–chromium–molybdenum–tungsten (CoCrMoW).

The major proportions of the elements used in cobalt–chromium alloys are cobalt (58–70%) and chromium (26–30%). The chromium constructs a secure oxide layer and protects the implant from generalized electrochemical dissolution.

In some alloys, such as ASTM F75 and F799, the addition of a small amount of molybdenum (5–7%) improves the passivation, which protects the implant from pitting attack from aggressive anions in the environment.

Nickel-containing alloys display an increase in friction and wear resistance but are not widely used due to cytotoxicity and allergic reactions to nickel when released into the body.

5.1.2.3. *Titanium and titanium alloys*

Pure titanium and its alloy (Ti6Al4V) are the most commonly used implant materials due to their high tensile strength, low density and easy fabrication. The most attractive characteristic of a titanium-alloy is that it has high corrosion resistance, since titanium readily forms a titanium dioxide (TiO_2) layer that forms a barrier to electrochemical reactions. The greater stability of the oxide layer on titanium alloys leads to their excellent biological compatibility.

Ti6Al4V is widely used in the manufacture of orthopedic implants. Aluminum (5.5–6.5%) and Vanadium (3.5–4.5%) are added to the titanium to stabilize the phases. Aluminum is an alpha phase stabilizer and Vanadium is a beta phase stabilizer. As a result, Ti6Al4V is an alpha-beta alloy. Their drawback is that titanium alloys have relatively high friction coefficients, however, that increase wear damage at the bearing surface in joint replacements.

5.1.2.4. *Overview*

In all the alloys described above, a stable oxide layer protects the underlying pure metal from continuous corrosion damage in a reactive environment. The corrosion damage is driven by electrochemical reactions. During exposure to aqueous conditions, the atoms at the metal surface experience an anodic process – electrons are released and metal ions are subjected to oxidation. The locally-established electrical potential accelerates localized corrosion damage.

In this chapter, the influences of the surface topographies of commercially available orthopedic implants that are critical with regards to service life, will be discussed to answer the primary question about how surface roughness determines the tribological (wear) and electrochemical (corrosion) properties of metallic implants. This chapter will provide a complete understanding of the mechanism of wear and corrosion damage and present desirable surface characteristics for use in the manufacture of implants. In this regard, this chapter deals with the issues of changes in surface roughness due to synergistic surface degradation during fatigue contact and physiological corrosion.

5.2. Clinical studies and mechanistic investigation into implant failure

Implant wear and corrosion is a complex multi-factorial degradation process that includes solution chemistry, metal oxidation, surface roughness and contact load. Metallic implant corrosion occurs due to a reaction with physiological environment if there is no tenacious surface oxide layer. Although the protective oxide layer does initially exist, when the surface layer becomes permeable to oxygen and moisture the electrochemical reaction will continue and lead to corrosion damage.

Implant retrieval studies have demonstrated serious corrosion damage at the modular junction of implant devices, as presented in Figure 5.2. The major damage was mostly due to wear at the articulating surfaces and significant fretting damage was found at the modular junctions [GIL 93, GIL 97, GIL 98, MCC 97, MCK 00]. For example, fretting wear and corrosion have been observed at the taper interface of the head–stem connection of a hip joint replacement [H24-36]. The implant retrieval analysis demonstrated that contact fatigue process produces particulate wear debris that initiates the formation of a granulomatous tissue that penetrates the bone–implant interface and results in aseptic loosening.

In the case of a dissimilar metal junction (Ti6Al4V on CoCrMo), the greater fretting damage was shown while the same metal combination results in less wear damage at modular connections [LIE 94]. *In vivo* fretting and passive dissolution influence osteoblasts by decreasing their proliferation, decreasing bone matrix formation and increasing interleukin-6 cytokine release [HAL 02]. The continuous change in physiological condition implies a local and systemic increase in toxicological events. Wear debris of high metallic composition are found in the surrounding tissue near the implantations and also in remote organs [URB 94, URB 00].

Migration of the soluble and particulate debris onto the vicinity results in inflammatory events in surrounding tissues and cells. The dissemination of wear debris that migrates to lymph nodes induces immune activation and the recruitment of inflammatory cytokines [HAL 03b].

Fretting wear and the corrosion response of modular implants has generally been quantified through electrochemical measurements of weight loss during fretting testing. The results of these studies indicated that material degradation was accelerated by the combination of mechanical loading and corrosion. Gilbert and Jacobs explained that there was an increased rate of corrosion due to the continual fracture and reformation of oxide layers (repassivation) over the metallic surfaces. Accordingly, a mechanical abrasion oxide layer is necessary to initiate increased electrochemical activity on the surface.

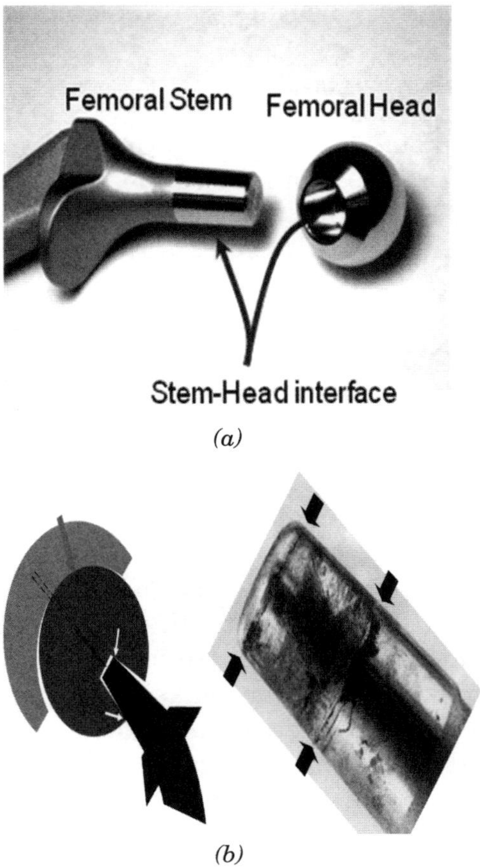

Figure 5.2. *(a) Modular interface of hip joint replacement; and (b) wear damage on a tapered stem surface*

The physiological fluid penetrates the implant's modular interface, leading to synergistic interactions of mechanical loading and electrochemical oxidation. The repeated contact loading of metallic implant interfaces results in the formation of soluble and particulate debris. The corrosion products from particulate debris may migrate locally and systemically and may induce an allergic response or inflammatory events. Those hostile reactions from human tissue ultimately generate bone loss and subsequent implant failure.

5.2.1. *Wear mechanisms*

The mechanism of wear is very complex and is combined with two or more types of material loss. The groups of wear mechanism are classified as adhesion, abrasion, erosion and fretting. Abrasive wear denotes scraping or cutting off of surface by a surface hard enough to produce more damage to a surface other than itself. Adhesive wear is usually found on a clean surface where high friction is established due to adhesion at areas of local contact. This mechanism involves separation of the adhesion points by tearing and fragmentation. In the case of cyclic contacts, however, the mechanism implies fatigue mode wear, such as fretting and delamination.

Implant surface damage is due to the fatigue mode wear damage process. The daily activities of patients apply cyclic contacts on the bearing surface and modular interfaces. The well-polished bearing surface maximizes the contact area and thus prevents local stress elevations. The secure oxide layers separate the two metal surfaces and avoid adhesion. The tapered surface on the femoral stem is brought into rough surface contact, so that the small contact areas will be subjected to local stress elevations and become plastically deformed. In addition, the relative displacement of two mating surfaces will break the oxide layer at contact points.

If the process includes repeated contact loading, however, the mechanism has to be engaged in fatigue mode damage. Most of the commercial materials contain multiple defects. Some defects are dislocations due to void, inclusion and atoms missing in crystalline structure. During repeated contact loadings, the shear load causes dislocation mobility and such dislocations accumulates at the subsurface.

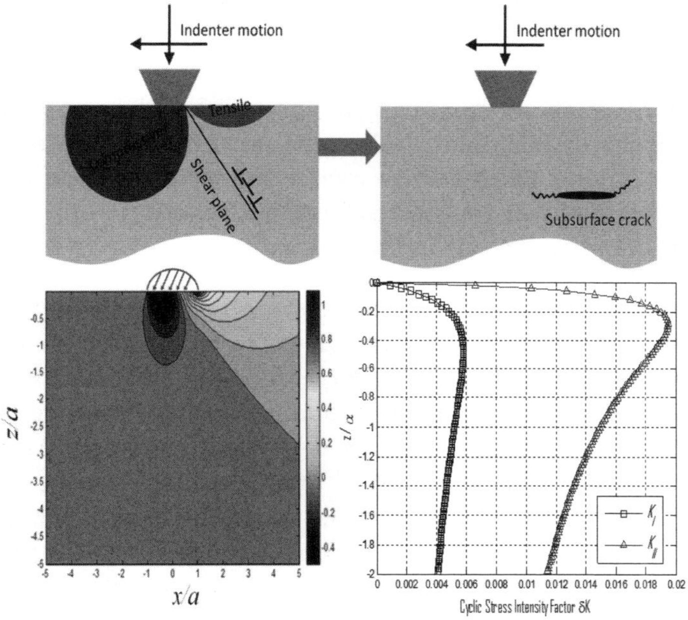

Figure 5.3. *The initiation of a subsurface crack by frictional contact stress and the stress intensity factor at the subsurface crack*

The dislocation pile-up creates micro-cracks which then coalesce and propagate (see Figure 5.3). Previous studies on the mechanism by which cracks are created below the surface due to a moving asperity, have illustrated that cracks propagate parallel to the surface due to the shear mechanism, while the cracking motion impacts the free surface due to a tensile mechanism. The location of the subsurface crack and friction coefficient are the major parameters involved in producing plate-shaped wear debris in ductile materials.

5.2.2. *Physiological corrosion of metals*

Metal corrosion is the process of degradation of metallic materials by reaction with their environment. Most metallic materials are susceptible to corrosive attack if they do not have a tenacious oxide layer on their surface. When the surface layer is permeable to oxygen and moisture, however, the process of corrosion will continue and lead to eventual failure. Among the various mechanisms of corrosion, metal corrosion is driven by electrochemical reactions – it includes a chemical reaction with electron transfer.

During exposure to an aqueous environment, atoms on the metal surface undergo an anodic process, where electrons are released, leading to oxidation and the formation of metallic ions. The simultaneous oxidation and reduction processes, as shown in Figure 5.4, produce the transformed metal ion M^{n+} and the free electron e^{n-}. The free electrons move through the conductive metal to other surface sites. An anodic process involves oxidation of metal to an ion with a valence charge and the release of electrons. The metal ion (M^{n+}) is released by an electrical potential and reacts with oxygen dissolved in the solution. The localized electrical potential accelerates the oxidation process until electrochemical potential is balanced.

The physiological environment is an oxygen-rich saline solution (containing about 0.9% salt) at an approximately neutral pH level of 7.4 [MUD 03]. The body fluid is electrochemically active due to the dissolved oxygen and reactive molecules that are present in the forms of sodium and chloride ions. Water, which is conductive, is the electrolyte.

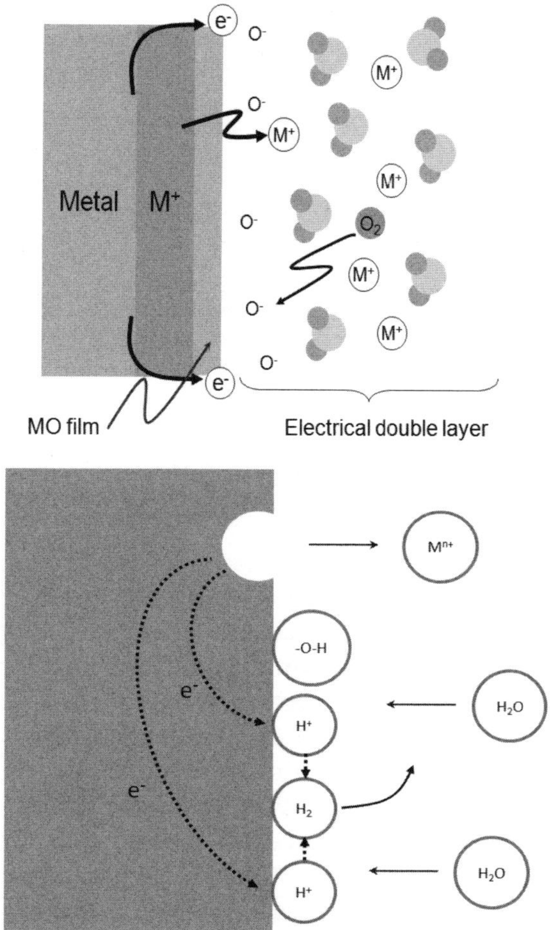

Figure 5.4. *Interface of a metal surface in contact with a biological environment and electrochemical dissolution*

In this corrosive physiological environment, contact loading of the rough surface probably induces a small-scale oxide film fracture leading to a local variance in its electrochemical potential. Preferential anodic reactions then produce localized surface defects and initiate microscale surface cracks. Repetitive loading on the implant material by human movement may propagate the surface crack. The synergistic combination of corrosion and surface layer fracture accelerate the damage occurring at the interacting surfaces.

Repetitive oxidation leads to a drop in pH level (due to an increase in local hydrogen concentration) whereby oxygen is consumed from the body fluid, thus increasing electrochemical reactivity. This event produces corrosion of products near the implant site and leads to problems involving biocompatibility.

5.2.3. *Bio-tribo-corrosion at the modular interface*

While ceramic materials have relatively high friction coefficients and high wear resistance, polymeric materials have low friction coefficients and low wear resistance. Metallic implants have moderate friction coefficients and suitable wear resistance. Contact fatigue occurring at the interface between two interacting surfaces, however, degrades the mechanical and chemical properties of the material used. Repeated friction leads to local plastic deformation, pitting and the appearance of scratches. Localized plastic deformation produces small-scale dislocations and the accumulated dislocations result in micro-cracks at the subsurface.

The surface damage induced by mechanical abrasion forms surface cracks that will propagate in subsequent interactions. The repetitive contact creates further plastic deformations, not only at the contact asperity but also at the crack tips that have already been produced. In long-range surface interactions, residual stresses that are created by plastic deformations due to repeated contact significantly reduce the fatigue strength of the material.

Modular orthopedic implant designs have been widely accepted as promoting optimized features for individual patients. The modular interface requires a certain degree of roughness to improve self-interlocking between the elements. When the surfaces are brought into contact, only the protruding surface asperities are in contact. The size and distribution of such points of contact depend on the roughness of the surface and the load that is applied. Even when the overall contact loads are not large enough to cause macroscopic deformation, plastic deformation can occur at a small scale under the contact asperities so that the surface roughness changes after removal of the contact load. The plastic deformation under asperity contact may lead to the development of a residual stress field at the surface of the material, which will in turn create stress-concentrations near the surface roughness grooves. Therefore, the surface roughness in modular implant design increases the severe cyclic relative motions (fretting) caused by the patient's activities. As a result, the interacting surface undergoes local plastic deformation and repetitive compressive and tensile tangential loadings act against the kinetic motion of the mating surfaces. The magnitude of loading is dependent on the normal loading and kinetic friction coefficient. Thus, interface roughness and cyclic loading will determine the contact fatigue damage process.

In the physiological environment, corrosive solution aggressively attacks areas of residual stress concentration, leading to the deepening of grooves and exacerbating surface deformation. The combined effect of plastic surface deformation and stress-assisted corrosion may ultimately result in surface damage by the nucleation of small cracks and breaking away of particulate debris. This mechanism is summarized in Figure 5.5.

Fretting motion at the implant interface causes the oxide layer fracture-reformation at the same time as the material is dissolved due to electrochemical reactions. The oxide layer, which separates the underlying metal from the reactive environment, is the primary reason for a metallic implant's corrosion resistance.

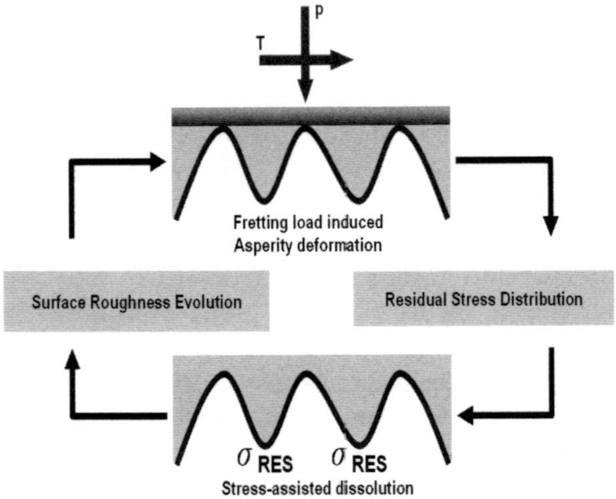

Figure 5.5. *Schematic of the mechanism of stress-assisted dissolution due to the combined influence of contact loading and corrosion*

The alloys chosen for implants are not perfectly inert in chemical reaction, but the native oxide film separates the metals from oxygen and moisture. Oxide film fracture exposes the pure metal to the corrosive environment that dissolves metal ions and builds an oxide layer during this period of surface damage. In the modular design of femoral components used in total hip joint replacements, the head–stem interface will have a metal-on-metal connection at the conical tapered surfaces.

A small gap between two surfaces will be created by diameter and angular tolerance in implant design. This gap is considered to be a crevice that can be penetrated by body fluid. Thus, modular tapers are susceptible to fretting contact and crevice corrosion.

Repassivation of the damaged oxide layer, which consumes oxygen in the solution, causes an increase in H^+, which lowers pH level within the crevice environment. The pH drop accelerates corrosion damage at the implant surface. The unstable oxidation will significantly modify the physical and mechanical properties of the metal's surface and corrupt wear resistance. At the same time, the crevice corrosion phenomenon constructs a strong oxidizing environment in the vicinity of the implant so that corrosion of the degraded surface will accelerate implant surface damage.

Local damage by material loss drastically reduces the wear resistance and fatigue strength of an implant. The soluble and particulate debris formed through the process may migrate locally and systematically and this can induce an adverse reaction within the patient's body.

5.2.4. *Adverse effects due to the creation of wear particles*

Corrosion is an undesirable deleterious process that occurs with metallic implants. The biological and chemical conditions (pH, solution protein, enzymes) in body fluid facilitates an aggressive corrosive environment. The electrochemical reaction is a chemical consequence of oxidation and reduction. The oxygen dissolved in the body's solution leads to an anodic process that removes metal ions from the surface of the implant. Fretting contact and crevice corrosion have been used to explain the implant surface damage. Oxidation in the small volume of modular space will limit electrochemistry and lead to unstable oxidation, known as crevice corrosion.

The biological response due to protein adsorption and cellular interaction with the implant surface causes various biocompatibility problems, such as inflammation, immune responses and blood clotting. The wear particles and dissolved metal ions, which are toxic, generate an inflammatory event in the vicinity of the implant. This is worsened by the fact that the inflammatory cells produce oxidizing agents that induce a strong corrosive environment. This inflammatory event ultimately leads to bone loss and the premature fracture of implants.

5.3. Residual stress development by rough surface contact

Implant materials are always subjected to stress, even when no external load is applied. The residually developed stress can be established during the manufacturing process, work-hardening, heat treatment and plastic deformation developed before insertion into a patient. The residual stress influences the material's properties. Tensile residual stress may cause early failure of materials during tensile testing because the tensile stress applied adds to the residual tensile stress. Residual stress therefore significantly modifies the mechanical properties of materials, such as fatigue strength.

Similarly, the modular interface between the femoral components obtained by male-female fixation undergoes fretting contact and develops a contact residual stress field due to human movement. During the useful life of a replacement joint, human behavior brings the head–stem interface into repetitive rough surface contact, leading to a non-uniform residual stress field. Plastic deformation during the contact loading–unloading process results in an elastic stress field after complete unloading due to the incorrect fit between the plastically deformed material and the recovering elastically deformed material. Once the contact stress of a material increases beyond the yield point, its stress state remains at yield point, while the circular plastic size expands. Finally the plastic deformation will reach the free surface and further contact pressures will deform the contact edge.

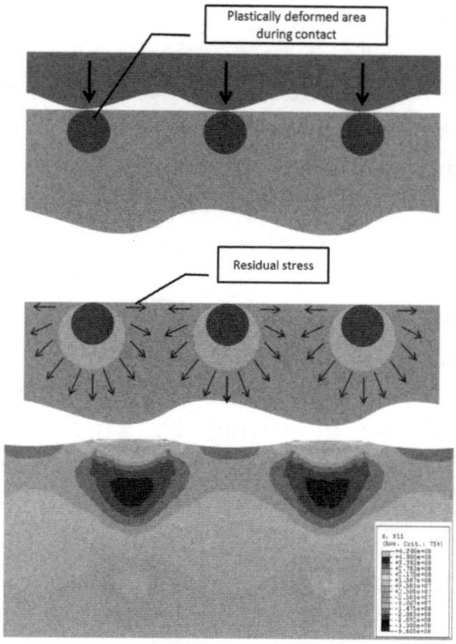

Figure 5.6. *Contact loading-induced residual stress on an undulating surface*

During contact loading, the elastic and plastic deformation depends on the surface contact configurations as well as the contact pressures, while the plastically-deformed region is responsible for the resultant residual stress field in the unloading process. In the two-dimensional (plane strain) contact problem, cylindrical plastic deformation is established by contact loading and during unloading of the plastically deformed area below contact asperity leads to elastic stress, which interacts with the adjacent elastic field induced by the neighboring plastically deformed area [HAR 71]. In modular implants, the rough surface at the head–stem junction establishes the integrity of modular implant components, while the nanoscale asperity contacts lead to localized plastic strains. Elastic strains at the asperities in contact developed after unloading generate stress concentrations at the surface troughs between neighboring asperity contacts.

The influence of surface roughness and residual contact stress on physiological corrosion has been investigated [RYU 07]. At the modular interface, rough surface contact nucleates the local plastic deformation of contact asperities and development of a residual stress field on unloading, as illustrated in Figure 5.6. Since residual stress due to contact depends on the roughness of the interfaces that are in contact and pressure they are under, the surface implant corrosion damage in an underlying corrosive physiological environment will be governed by surface roughness and contact loads. The finishing processes used in modular implant design lead to an undulating surface with certain degrees of roughness, with the amplitude and wave-number controlled by the tool tip size and cutting speed. During the service life of a hip joint, human behavior brings the head–stem interface into repetitive rough surface contact, leading to a non-uniform residual stress field. Stress concentration due to rough surface contact induces an acceleration in the creation of localized corrosion pits. The thermodynamic driving force that constitutes electrochemical dissolution depends on the surface stress established by fatigue contact. Localized corrosion damage at the implant surface is due to the spatial variation in elastic energy induced by non-uniform residual stresses.

5.3.1. *Surface properties of bio-implants*

Surface properties involve the physical and chemical responses of a surface in a physiological environment. Physical properties include the surface roughness and surface tension that governs cell adsorption. At the taper junction of modular implants, the surface is usually manufactured by a machining process to an undulating profile to promote interlocking. In this rough surface contact, hardness is an important mechanical property. It characterizes resistance to penetration by indentation because the hardness of the material is often equated with wear resistance and durability. In metallic materials it serves as a measure of abrasion resistance and strength. Ceramic-on-metal implants have a lower wear rate at the interface than metal-on-metal pairings, since the hardness of ceramic surfaces is much greater than that of metals.

The repassivation of metal oxide lowers wear damage. The damaged metallic implant surface rapidly reconstructs an oxide layer and protects against further exposure to oxygen and moisture. In the reactive environment, the mechanism of wear may change due to more aggressive chemo-mechanical reactions and, as a result, more material may be removed, even in a similar range of contact pressures. Accordingly, the wear mechanism is dependent on the competition between oxidation, which produces a protective layer, and electrochemical dissolution.

A typical modular junction interface has an amplitude-to-wavelength ratio of 0.003 to 0.005 (see Figure 5.7). Its mechanical properties have been improved by a variety of technologies involving the compatible matching of materials, metallurgical surface property enhancement and implementation of a surface layer. Other innovative technologies have been employed to enhance the reliability and useful life of a replacement joint by obtaining a stable state due to joining surface geometrical modifications, adding porous coating, surface texturing, etc. Chemical properties are related to a joint's surface electrochemical response in solution.

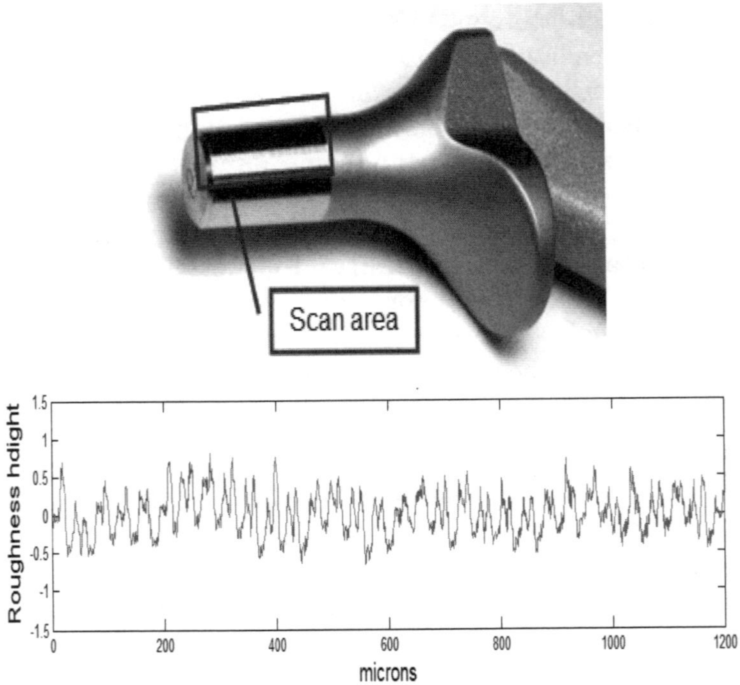

Figure 5.7. *Surface roughness of a tapered stem*

5.3.2. *Stress-assisted electrochemical dissolution and local corrosion damage*

The fatigue strength of metal alloys is affected by being immersed in an aqueous environment. The basic mechanism of fatigue strength degradation in an aqueous environment is related to fatigue crack initiation. The anodic process creates local corrosion pits and leads to areas of stress concentration, which accelerates corrosion damages. The mechanical load opens and shears the crack produced by a corrosion pit and the oxidation prevents the reversal of the creation of slip bands in small plastic strains near the crack tips.

This synergistic effect results in an increase in stress level and crack growth rate. During contact fatigue at the surface of metal alloys in a corrosive environment, the stress concentration will occur at surface recessions due to the residual stress established by plastic deformations at the surface asperities. The electrochemical potential of a strained surface increases and accelerates dissolution. Measurement of the roughness evolution at nanoscale indicates that the dissolution of materials is concentrated in troughs on the surface.

This process increases the surface roughness and ultimately creates surface cracks. During fatigue contact, the crack propagates in two different manners. Bending stress on the stem may lead crack propagation into the material; while fatigue contact stress may lead to crack propagation parallel to the surface.

5.3.2.1. *Thermodynamic driving forces and roughness evolution*

Electrochemical dissolution of a stressed surface via chemical reactions is due to the thermodynamic forces that cause instability of the solid–liquid interface. The reaction rate at the solid surface is determined by the surface mobility (*M*) and the driving force (*F*). The surface mobility will vary when the reaction responsible is thermally stimulated. The driving force mainly consists of chemical potential, strain energy and interfacial energy. The total electrochemical reaction rate is characterized by the linear superposition of the driving forces such as;

$$R = MF$$

$$F = g - U_{SE} - \gamma K$$

where:

$-R$ is the chemical reaction rate;

$-M$ is the mobility of solid surface in liquid;

– g is the chemical potential;

– U_{SE} is elastic strain energy;

– γ and K are surface energy and curvature, respectively.

Chemical potential (g) will be varied by the physical condition of host's tissue.

If the chemical potential is consistent over the surface and the surface is stressed by mechanical contact, however, the total material dissolution rate includes uniform dissolution by consistent chemical potential and non-uniform dissolution by local deviations of elastic strain energy (U_{SE}) and surface curvature (K). Uniform dissolution is responsible for the average changes in surface height during chemical exposure, while the non-uniform dissolution leads to changes in roughness. At the implant surface, damage due to corrosion is susceptible to elastic strain energy via non-uniform plastic contact deformations and surface curvature (roughness).

The cyclic contact of an undulating surface manufactured by machining, results in high stress variations due to plastic deformation of the asperities in contact. Even after complete unloading, the imperfect fit between plastically-deformed materials and the surrounding elastically-deformed materials establishes a residual local stress concentration at the troughs on the surface.

Stress-dependent relative dissolution occurs via a chemical reaction, as illustrated in the thermodynamic driving force equation. The more aggressive corrosion takes place at highly stressed surface troughs rather than at the protruding surface area, so that the greater corrosion damage at surface grooves is associated with an increase in surface roughness (see Figure 5.8). The complex mechanism of implant corrosion may therefore be summarized by differential material dissolutions between asperity summits and surface troughs, which can be determined by contact load and surface roughness characteristics. During the service life of implants, the roughening process continues and initiates cracks in the recessed surface area. This constitutional relation characterizes the onset of corrosion-fatigue damage in orthopedic implants.

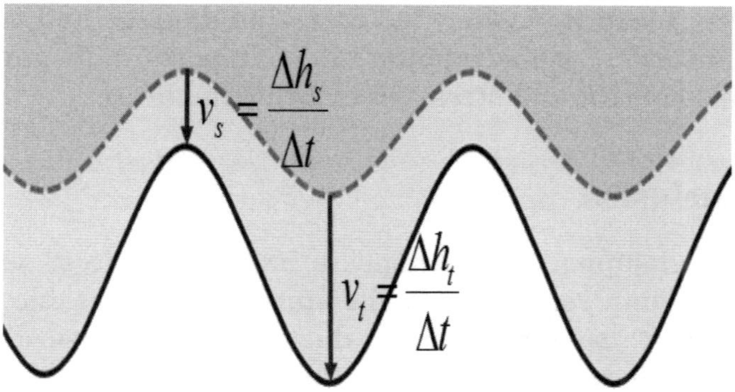

Figure 5.8. *Roughness evolution and localized surface damage due to stressed surface corrosion*

5.3.2.2. *Influence of residual stress on roughness evolution in fatigue contact and environmental corrosion*

The evolution of the roughness of a stressed surface during a corrosion experiment showed an increase in the amplitude of particular roughness modes [KIM 99, SAG 00]. In the simulation of joint replacement corrosion, the contact-stressed implant material was subjected to an increase in surface roughness at the most dominant roughness components [RYU 07]. This phenomenon depends on surface roughness, the level of contact loading and yield strength. These results suggest that the magnitude, as well as the distribution of residual stress fields induced, due to the plastic deformation asperities in contact, influence material dissolution during corrosion. Therefore corrosion damage at implant surfaces is associated with material dissolution, which will be determined by the free energy change constituted by local stress states. The evolution of a particular group of surface roughness modes during environmental corrosion again results in the roughening of an implant's surface and, consequently, increased surface amplitude, increasing the residual contact stress. The synergistic interaction of residual stress and stress-assisted dissolution lead to accelerated corrosion damage and surface crack initiation. Consequently, it is important to suppress the effects of residual stress on metallic implants.

5.4. Conclusion

For load-supporting orthopedic implant devices, such as total hip joint replacement, in spite of their stable oxide layer (inert surface), high ductility and wear/friction resistance, their mechanically-assisted corrosion in biological fluids degrades the surfaces' chemical and mechanical properties. The complex mechanism of damage was investigated via mechanical and chemical analyses of the plastic deformations of metallic implants during patient activities. It was found that there was a synergistic interaction between the increase in local corrosion pits and degradation of fatigue strength and wear resistance. Currently, manufactured modular implant surfaces usually have periodically scattered protruding asperities and recessed troughs in order to enhance desired interlocking between the modular components.

The electrochemical reactivity of a solid surface interfacing with a reactive solution is driven by surface roughness, interfacial energy, strain energy and chemical potential. On the bearing surface, delamination wear occurs due to fatigue sliding contact. Among the multi-factorial combinations, contact-induced residual stress is a predominant factor influencing the chemo-mechanical phenomena leading to physiological corrosion damage.

Fretting contact at the rough surface results in a high variation on residual stress along the interface and stress-assisted preferential dissolution leads to localized surface damage – so-called fretting corrosion. The wide range of fretting corrosion attack on implant surface nucleates surface cracks that amplify the surface stress concentrations. Identification of the transition from the evolution of surface roughness to surface crack initiation or local pitting corrosion is very important. This is needed to characterize the onset of corrosion damage. The steady stage of roughness evolution, however, only describes the threshold period of an implant's useful life.

Surface roughness and contact loading of metallic implant materials are therefore the primary factors to be considered in interface design. Although corrosion damage in implants cannot be measured or predicted, wear and corrosion damage can be mitigated if a well-established surface design increases the contact area and suppresses plastic strains. Secure oxide layer growth will enhance protection against corrosion. The surface stress that drives oxidation may be manipulated to produce a sufficiently thick, stable protective film.

5.5. Bibliography

[AZI 91] AZIZ M.J., SABIN P.C., LU G.Q., "The activation strain tensor – nonhydrostatic stress effects on crystal-growth kinetics", *Physical Review B*, vol. 44, p. 9812-9816, 1991.

[BAR 98] BARVOSA-CARTER W., AZIZ M.J., GRAY L.J., KAPLAN T., "Kinetically driven growth instability in stressed solids", *Physical Review Letters*, vol. 81 p. 1445-1448, 1998.

[GIL 93] GILBERT J.L., BUCKLEY C.A., JACOBS J.J., "In vivo corrosion of modular hip prosthesis components in mixed and similar metal combinations. The effect of crevice, stress, motion, and alloy coupling", *J. Biomed Mat Res*, vol. 27, p. 1533-1544, 1993.

[GIL 97] GILBERT J.L., JACOBS J.J., "The mechanical and electrochemical processes associated with taper fretting crevice corrosion: A review", in: MARLOWE D.E., PARR J.E., MAYOR M.B. (eds.), *Modularity of Orthopedic Implants, American Society Testing and Materials*, W. Conshohocken, p. 45-59, 1997.

[GIL 98] GILBERT J.L., "Step-polarization impedance spectroscopy of implant alloys in physiologic solutions", *Journal of Biomedical Materials Research*, vol. 40, p. 233-243, 1998.

[HAL 02] HALLAB N.J., VERMES C., MESSINA C., ROEBUCK K.A., GLANT T.T., JACOBS J.J., "Concentration- and composition-dependent effects of metal ions on human MG-63 osteoblasts", *Journal of Biomedical Materials Research*, vol. 60, p. 420-433, 2002.

[HAL 03a] HALLAB N.J., JACOBS J.J., "Orthopedic implant fretting corrosion", *Corrosion Reviews* , vol. 21, p. 183-213, 2003.

[HAL 03b] HALLAB N.J., CUNNINGHAM B.W., JACOBS J.J., "Spinal implant debris-induced osteolysis", *Spine*, vol. 28, p. S125-S138, 2003.

[HAR 71] HARDY C., BARONET C.N., TORDION G.V., "The elasto-plastic indentation of a half-space by a rigid sphere", *International Journal for Numerical Methods in Engineering*, vol. 3, p. 451, 1971.

[JAC 98] JACOBS J.J., GILBERT J.L., URBAN R.M., "Corrosion of metal orthopaedic implants", *Journal of Bone and Joint Surgery American*, vol. 80A, p. 268-282, 1998.

[JAC 03] JACOBS J.J., HALLAB N.J., SKIPOR A.K., URBAN R.M., "Metal degradation products – A cause for concern in metal-metal bearings?", *Clinical Orthopaedics and Related Research* p. 139-147, 2003.

[KIM 99] KIM K.S., HURTADO J.A., TAN H., "Evolution of a surface-roughness spectrum caused by stress in nanometer-scale chemical etching", *Physical Review Letters*, vol. 83, p. 3872-3875, 1999.

[LIE 94] LIEBERMAN J.R., RIMNAC C.M., GARVIN K.L., KLEIN R.F., SALVATI E.A., “An analysis of the head-neck taper interface in retrieved hip prostheses”, *Clinical Orthopaedics*, vol. 300, p. 162, 1994.

[MCC 97] MCCARTHY J.C., BONO J.V., ODONNELL P.J., “Custom and modular components in primary total hip replacement”, *Clinical Orthopaedics and Related Research*, p. 162-171 1997.

[MCK 00] MCKELLOP H., SHEN F.W., LU B., CAMPBELL P., SALOVEY R., “Effect of sterilization method and other modifications on the wear resistance of acetabular cups made of ultra-high molecular weight polyethylene – A hip-simulator study”, *Journal of Bone and Joint Surgery – American*, vol. 82A, p. 1708-1725, 2000.

[MIT 08] MITCHELL A., SHROTRIYA P., “Stress assisted dissolution of biomedical grade CoCrMo: Influence of contact loads and residual stresses”, *Acta Biomaterialia*, vol. 4, p. 296-304, 2008.

[MUD 03] MUDALI U.K., SRIDHAR T.M., RAJ B., “Corrosion of bio implant”, *Sadhana*, vol. 28, p. 601-637, 2003.

[RYU 07] RYU J., DAYAL V., SHROTRIYA P., “Onset of surface damage in modular orthopedic implants: Influence of normal contact loading and stress-assisted dissolution”, *Experimental Mechanics*, vol. 47, p. 395, 2007.

[SAG 00] SAGE J., BARVOSA-CARTER W., AZIZ M., “Morphological instability of growth fronts due to stress-induced mobility variations”, *Applied Physics Letters*, vol. 77, p. 516-518, 2000.

[SUH 73] SUH N.P., “The delamination theory of wear”, *Wear*, vol. 25, p. 111-124, 1973.

[SUH 74] JAHANMIR S., SUH N.P., ABRAHAMSON E.P., “Microscopic Observations of the Wear Sheet Formation by Delamination”, *Wear*, vol. 28, p. 235-249, 1974.

[TEM 08] TEMENOFF J.S., MIKOS A.G., *Biomaterials: The Intersection of Biology and Materials Science*, Pearson-Prentice Hall, 2008.

[URB 94] URBAN R.M., JACOBS J.J., GILBERT J.L., GALANTE J.O., "Migration of corrosion products from modular hip prostheses – particle microanalysis and histopathological findings", *Journal of Bone and Joint Surgery – American*, vol. 76A, p. 1345-1359 1994.

[URB 00] URBAN R.M., JACOBS J.J., TOMLINSON M.J., GAVRILOVIC J., BLACK J., PEOC'H M., "Dissemination of wear particles to the liver, spleen, and abdominal lymph nodes of patients with hip or knee replacement", *Journal of Bone and Joint Surgery – American*, vol. 82A, p. 457-477, 2000.

[VIC 96] VICECONTI M. *et al.*, "Design-related fretting wear in modular neck hip prosthesis", *Journal of Biomedical Materials Research*, vol. 30, p. 81-186, 1996.

[VIC 97] VICECONTI M., BALEANI M., SQUARZONI S., TONI A., "Fretting wear in a modular neck hip prosthesis", *Journal of Biomedical Materials Research*, vol. 35, p. 207-216, 1997.

[WIL 76] WILLIAMS D.F., "Corrosion of implant materials", *Annual Review of Materials Science*, vol. 6, p. 237-266, 1976.

[WIL 93] WILLMANN G., "Fretting of modular implant systems", *Biomedizinische Technik*, vol. 38, p. 48-52, 1993.

List of Authors

Jayashree BIJWE
Industrial Tribology Machine Dynamics & Maintenance
Engineering Center Indian Institute of Technology
Delhi
India

J. Paulo DAVIM
Department of Mechanical Engineering
University of Aveiro
Portugal

Xue GONG
State Key Laboratory of Solid Lubrication
Lanzhou Institute of Chemical Physics
China

Jiesheng HAN
State Key Laboratory of Solid Lubrication
Lanzhou Institute of Chemical Physics
China

Xianguo HU
Institute of Tribology
Hefei University of Technology
China

Zhen LI
State Key Laboratory of Solid Lubrication
Lanzhou Institute of Chemical Physics
China

Jinjun LU
State Key Laboratory of Solid Lubrication
Lanzhou Institute of Chemical Physics
China

Junhu MENG
State Key Laboratory of Solid Lubrication
Lanzhou Institute of Chemical Physics
China

Jae-Joong RYU
Mechanical and Industrial Engineering Department
Texas A&M University
Kingsville
USA

Mohit SHARMA
Leibniz Institute of Polymer Research
Dresden
Germany

Pranav SHROTRIYA
Mechanical Engineering Department
Iowa State University
Ames
USA

Yufu XU
Institute of Tribology
Hefei University of Technology
China

Hu WANG
Institute of Tribology
Hefei University of Technology
China

Hengzhou WO
Institute of Tribology
Hefei University of Technology
China

Belal F. YOUSIF
Faculty of Engineering and Surveying
University of Southern Queensland
Queensland
Australia

Index

A, B

abrasive wear, 6, 24, 29, 38
adhesive wear, 14, 21, 29, 45, 61, 65, 66, 78, 94, 95
advanced ceramic, 112
area fraction of cavities, 103, 104
back pressure, 17, 37
bio-implants, 186
bio-tribo-corrosion, 180

C

carbon
  fabric reinforced composites (CFRCs), 12-16, 19, 29, 30, 41, 42, 44, 46, 48, 54, 57
  fibers, 2, 3, 6-11, 15, 20, 27-30, 32-35, 37, 38, 40, 42, 44, 46-48, 57, 58
cast iron, 107-111, 113
cavitation, 1-7, 10-19, 21-26, 28, 30, 32-35, 37, 38
  erosion, 1-5, 10, 11, 13, 14, 28, 30, 32-34, 38, 99-106, 108, 109, 113-116
  flow, 1, 13, 14, 16, 37
  inception, 16, 26, 28, 38
clinical studies, 173
cold remote nitrogen-oxygen plasma (CRNOP), 34, 37, 38, 48, 49
composite development, 38
contact conditions, 62, 66, 76-78, 80, 82, 83, 85, 87, 90-95
copper alloy, 109
corrosion, 3, 11-13, 36-38
  damage, 165, 169, 173, 174, 183, 186, 188, 190-193
  product, 166, 168, 176
  resistance, 108, 110, 112, 115, 116
cumulative mean depth of erosion, 103, 104, 106

D, E

diesel engine, 1
different shapes, 19
dimensionless parameters, 24
environmental corrosion, 191, 192
erosion of nozzle, 36
erosive wear, 28, 47

F

fabric
  orientation effect, 29
  weaves, 12
fatigue contact, 164, 173, 186, 189, 191
fiber
  matrix interface, 33
  precursor, 8, 9
  properties, 9
film technique, 17, 18, 30, 32, 47, 55
flow characteristics, 22, 23, 38
fretting wear, 2, 26, 47
fuel property, 21

H, I

hand lay up technique, 20
hardness of surface, 33
heat treatment, 9-11
hydraulic characteristics, 13
implant failure, 164, 173, 176
impregnation technique, 16, 18-20, 28, 30, 32, 42, 55
incubation time, 100, 103, 104, 113, 116
injection
  nozzle, 28, 37, 38
  pressure, 13, 14, 16, 17, 22-25, 37, 38
inlet radius, 13, 18
interfacial adhesion, 66, 69-72, 74, 84, 87, 91

L, M

local corrosion damage, 188
manufacturing methods, 13
material selection, 99, 101, 108, 115
maximum erosion rate, 103, 104
mechanical properties, 62, 65, 74
mechanism of cavitation erosion, 4, 5, 11
metal alloys, 165, 169, 170, 188, 189
micro-jet formation theory, 8, 9
modular interface, 164, 168, 175, 176, 180, 181, 184, 186
molecular weight of a polymer, 42, 45

N, P

nano-particles, 34
NaOH treatment, 69, 87
natural fibers, 62-65, 67
nickel alloy, 109
nitric acid treatment, 34, 36, 49
nozzle, 1-4, 11, 13-20, 26, 28-33, 36-38
  geometry, 18
numerical models, 15
physiological corrosion of metals, 178
polyester composites, 61, 65-67, 78
polymer, 107, 113
polymeric
  composites, 62-65
  tribo-composites, 3

R, S

rare earth compound, 34
residual stress, 163, 180, 181, 184-186, 189, 191-193

rough surface contact, 168, 176, 184, 186
roughness evolution, 189, 191, 193
stainless steel alloys, 170, 171
steel, 102, 105, 107-109, 111, 112, 116
strengthening, 33, 49
stress-assisted electrochemical dissolution, 188
surface
  properties, 169, 186
  roughness, 103, 104, 108

T, U

thermal effect, 10
thermodynamic driving forces, 189
total joint replacements, 167
tribological
  properties, 2, 14, 33, 38, 47, 49
  studies, 38, 45
ultrasonic vibratory apparatus, 101-104, 114-116

V, W

validation, 15
vibratory cavitation erosion apparatus, 99, 101
wear, 163-166, 168-170, 172-176, 178, 180, 183, 187, 192, 193
  mechanisms, 176
  particles, 168, 183
  resistance, 99-101, 103-110, 112-116
weave pattern, 12, 13, 54
worn surfaces, 65, 79, 84, 85, 87, 90